SpringerBriefs in Complexity

For further volumes:
http://www.springer.com/series/8907

Springer Complexity

Springer Complexity is an interdisciplinary program publishing the best research and academic-level teaching on both fundamental and applied aspects of complex systems—cutting across all traditional disciplines of the natural and life sciences, engineering, economics, medicine, neuroscience, social and computer science.

Complex Systems are systems that comprise many interacting parts with the ability to generate a new quality of macroscopic collective behavior the manifestations of which are the spontaneous formation of distinctive temporal, spatial or functional structures. Models of such systems can be successfully mapped onto quite diverse "real-life" situations like the climate, the coherent emission of light from lasers, chemical reaction-diffusion systems, biological cellular networks, the dynamics of stock markets and of the internet, earthquake statistics and prediction, freeway traffic, the human brain, or the formation of opinions in social systems, to name just some of the popular applications.

Although their scope and methodologies overlap somewhat, one can distinguish the following main concepts and tools: self-organization, nonlinear dynamics, synergetics, turbulence, dynamical systems, catastrophes, instabilities, stochastic processes, chaos, graphs and networks, cellular automata, adaptive systems, genetic algorithms and computational intelligence.

The three major book publication platforms of the Springer Complexity program are the monograph series "Understanding Complex Systems" focusing on the various applications of complexity, the "Springer Series in Synergetics", which is devoted to the quantitative theoretical and methodological foundations, and the "SpringerBreifs in Complexity" which are concise and topical working reports, case-studies, surveys, essays and lecture notes of relevance to the field. In addition to the books in these two core series, the program also incorporates individual titles ranging from textbooks to major reference works.

Klaus Mainzer · Leon Chua

The Universe as Automaton

From Simplicity and Symmetry
to Complexity

 Springer

Prof. Dr. Klaus Mainzer
Technische Universität München
Lehrstuhl für Philosophie und
  Wissenschaftstheorie
Arcisstrasse 21
80333 München
Germany
e-mail: mainzer@cvl-a.tum.de

Prof. Dr. Leon Chua
EECS Department
University of California
Cory Hall #1770 253
Berkeley
CA 94720-1770
USA
e-mail: chua@eecs.berkeley.edu

ISSN 2191-5326
ISBN 978-3-642-23476-7
DOI 10.1007/978-3-642-23477-4
Springer Heidelberg Dordrecht London New York

e-ISSN 2191-5334
e-ISBN 978-3-642-23477-4

Library of Congress Control Number: 2011936531

Cover design: eStudio Calamar, Berlin/Figueres

Printed on acid-free paper

Springer is part of Springer Science+Business Media (www.springer.com)

Preface

This booklet is an essay at the interface of philosophy and complexity research, trying to inspire the reader with new ideas and new conceptual developments in cellular automata (CA). Although the text is introductory, it goes beyond the presentation of nice pictures with pattern formations. Steven Wolfram declared computer experiments with pattern formation by CA as a "new kind of science". We claim that even in the future, quasi-empirical computer experiments are not sufficient. CA must be considered complex dynamical systems in the strictly mathematical sense, with corresponding equations and proofs. In short, we need analytical models for CA, to find precise answers and predictions in the universe of CA. In this sense, our booklet goes beyond Wolfram's approach.

After a historical and philosophical introduction to the old question "Is the universe a (cellular) automaton?" CA are defined as complex dynamical systems. The geometrical representation of the eight CA-rules as a Boolean cube allows precise definitions of a complexity index and universal symmetries. It can be proved that the 256 one-dimensional CA are classified by local and global symmetry classes for CA. There is an exceptional symmetry group with universal computability which we call the "holy grail" in the universe of CA. Although the four automata of this group are completely deterministic, their long-term behavior cannot be predicted in principle with respect to the undecidability of Turing's famous halting problem. Many analytical concepts of complexity research (such as attractors, basins of attractors, time series, power spectra, and fractality) are defined for CA. But there are also surprising phenomena in the CA-world (isles of Eden) without analytical representation in dynamical systems.

Finally we ask whether CA can be considered models of the real world? We introduce a test procedure to decide between an arrow of time or time reversibility in the attractor dynamics of CA. Can we compare symmetries of the physical universe with symmetries in the toy world of CA? What are the similarities and differences? According to a famous hint by Richard Feynman, classical probabilistic and nondeterministic automata are not sufficient to simulate the quantum universe. Therefore, quantum CA are a promising field for future research. CA can also be considered models of complex networks in the life sciences and technology. We discuss applications in systems biology, brain research, and robotics. Self-organization and the emergence of structure and patterns can be made precise in the CA-context.

In the end it is not essential whether the universe is an automaton in some metaphysical sense. In any case, CA are beautiful and fascinating examples of a general tendency in modern research: The world is increasingly represented by digitized models to handle the increasing complexity of research by high-speed computers. The final philosophical question arises of whether the digitization of the world has limitations.

Both authors want to thank the Institute for Advanced Study (IAS) at the Technische Universität München (TUM), especially its director Patrick Dewilde, for supporting and enabling our cooperation in Munich. Leon Chua would like to thank the USA Guggenheim Foundation, the UK Leverhulme Trust Visiting Professorship, and the AFOSR grant no. FA9550-10-1-0290 for their generous supports. We also thank Christian Caron (Springer) and external referees for helpful hints.

Munich and Berkeley, April 2011   Klaus Mainzer and Leon Chua

Contents

Chapter 1
Introduction: Leibniz, Turing, Zuse, and Beyond

According to Einstein, a scientific explanation should be as simple as possible, but not too simple, for it to be realistic. It would be nice to understand the great scientific problems of the universe (such as cosmic expansion, black holes, the evolution of life, and brains) with just basic knowledge. The toy world of cellular automata is an intuitive, but mathematically precise model that may be used to illustrate fundamental problems of topical research. The philosopher and mathematician Gottfried Wilhelm Leibniz (1646–1716), who constructed one of the first mechanical calculating machines, considered, even then, the universe as an automaton created by God as a divine engineer and mathematician. The theory of cellular automata was independently initiated by several computer pioneers, among them John von Neumann (1903–1957) and Konrad Zuse (1910–1995).

Does nature behave like a computer? Describing the universe by means of bits and qubits (quantum bits) reveals new and surprising connections. Impulses in the relay chains of his first computers led Konrad Zuse to think of light quanta and pose the question: "What if, in principle, anything, no matter its size, can be understood in terms of quantum particles?" Starting from Zuse's visionary idea of a "Calculating Space" (Zuse 1969), deep philosophical questions arise: Does the universe calculate? Is there a program that controls the world? How can the self-organization of complex structures be generated by simple rules?

Leibniz' Vision of Divine Automata

Leibniz already had the idea of elementary automata representing the universe, and which he called *monads*. In his view, they are simple systems changing their states according to certain rules. The elementary automata constitute aggregations of more or less complex systems, which are characterized by different correlations. They constitute composite automata. Examples of these natural automata are cells, organs, or organisms. Thus, for Leibniz, the complexity of the world is mirrored in a monadic network of automata (Mainzer 1994). He was convinced that natural automata surpass all kinds of technical automata in an infinite number of ways.

K. Mainzer and L. Chua, *The Universe as Automaton*, SpringerBriefs in Complexity, DOI: 10.1007/978-3-642-23477-4_1, © The Author(s) 2012

Leibniz was also the inventor of binary numbers. They are the simplest way to represent numbers which, for Pythagorean philosophers, are the codes that represent the world. Thus, Leibniz strongly believed that God created a digital world of automata with states 0 and 1.

At the end of the seventeenth century, Leibniz' vision of a digitized world of automata was extraordinary. He also proclaimed a universal methodology of formal procedures (*mathesis universalis*) to solve any problem by mechanical calculations (algorithms). Actually, problems should be coded by symbols and numbers as in algebra or Leibniz' calculus, to implement them on mechanical calculating machines. Leibniz distinguished two basic disciplines of his *mathesis universalis*. An *ars iudicandi* should allow every scientific problem to be decided by an appropriate arithmetic algorithm after its codification into numeric symbols. An *ars inveniendi* should allow scientists to seek and enumerate possible solutions of scientific problems. Leibniz' *mathesis universalis* already seems to foreshadow the famous *Hilbert program* at the beginning of the twentieth century, with its demands for formalization and axiomatization of mathematical knowledge. In accordance with his mathematical philosophy of rationality, Leibniz was deeply convinced that there are universal algorithms able to decide all the problems in the world by mechanical devices.

Leibniz' historical calculation machine became the prototype of *hand calculating machines*. But each step of the calculation still had to be implemented by hand. At the beginning of the nineteenth century, it was the English mathematician Charles Babbage who constructed the first *program-controlled calculation machine* (the *Analytical Engine*). His programming technology was inspired by the beginning automation of looms with punch cards in the age of early industrialization. His smart co-worker, Lady Ada Lovelace, daughter of the famous poet Lord Byron, described Babbage's Analytical Engine as suitable for "developing and tabulating any function whatever... the engine [is] the material expression of any indefinite function of any degree of generality and complexity" (Toole 1992). She described its use for scientific computing, including trigonometric functions and Bernoulli numbers. She also championed the idea that it could be applied to produce music and graphics. Thus, for the first time, she articulated the idea of a general-purpose computer.

General-purpose computers are capable of running different algorithms without requiring any hardware modifications. This is possible because modern-day microprocessors are based upon the von Neumann architecture. In this model, computer programs and data are stored in the same main memory. This means that the contents of memory may be treated as a machine instruction or data, depending on the context.

The Church–Turing Thesis

But how can we define computational procedures independently of any technical standards for hardware? In 1936, Alan Turing introduced his famous logical-mathematical concept of a machine, which was intuitively inspired by a typewriter at that time, but formally independent of any technical demands (Turing 1936/1937). A Turing machine (TM) can carry out any effective procedure (algorithm) provided it is correctly programmed. An effective procedure manipulates symbols by simple instructions step by step.

Therefore, the program of a TM (a *Turing program*) models a machine that mechanically operates on a tape. On this tape are symbols, which the machine can read and write, one at a time, using a tape head. An operation is fully determined by a finite set of elementary instructions such as "in state 23, if the symbol seen is 0, write a 1; if the symbol seen is 1, shift to the right, and change into state 12; in state 12, if the symbol seen is 0, write a 1 and change to state 7;" etc. In general, a TM consists of

(a) a potentially infinite tape, divided lengthwise into squares,
(b) a head that can read and write symbols on the tape and move the tape left and right one (and only one) cell at a time,
(c) a finite table of instructions that, given the state the machine is currently in and the symbol it is reading on the tape, tells the machine to do the following in sequence: Either erase or write a symbol, and then move the tape, and then assume the same or a new state as prescribed,
(d) a state register that stores the state of the Turing table, one of finitely many. There is one special start state with which the state register is initialized.

If the symbols used by a TM are restricted to a stroke | and a blank $\star$, then a computable arithmetic function can be computed by a TM. Notice that every natural number x can be represented by a sequence of x strokes (for instance 3 by |||), each stroke on a square of the Turing tape. The blank $\star$ is used to denote that the square is empty (or the corresponding number is zero). In particular, a blank is necessary to separate sequences of strokes representing numbers. Thus, a TM computing a function f with arguments $x_1, ..., x_n$ starts with tape $...\star x_1 \star ...\star x_n \star...$ and stops with $...\star x_1 \star ...\star x_n \star f(x_1, ..., x_n) \star...$ on the tape (Mainzer 2003).

From a logical point of view, a general purpose computer is a technical realization of a *universal TM* (UTM) which can simulate any kind of Turing program (Herken 1995). A UTM takes as input the description of a TM (along with the initial tape contents of the TM), and simulates the input on that TM. Thus, the UTM can simulate the behavior of any individual TM. In modern terms, the UTM is an interpreter program performing a step-by-step simulation of any TM. Program and data are the same thing, because a program (Turing program) is nothing more than a sequence of symbols that looks like any other data. When input to a UTM, the program begins to compute. Obviously, Alan Turing's work on UTM anticipated the development of general-purpose computers. He described the first such TM in his 1936 paper. Claude Shannon proved that two symbols were sufficient, provided enough states were used (Shannon 1956). In 1962, Marvin Minsky discovered a remarkable 7-state UTM that executes using a four symbol alphabet, but it is rather complicated to describe (Minsky 1962).

In 1936, Alonzo Church and Alan Turing independently proposed models of computation that they believed realized the notion of computation by a mechanistic procedure. Church invented the *lambda calculus* to study notions of computability while Turing used his TMs. Although both models appear very different from one another, Turing later showed that they were equivalent in that they each pick out the same set of mathematical functions (Turing 1937). There are many other mathematically equivalent procedures for defining computable functions. *Recursive functions* are defined by procedures for functional substitution and iteration, beginning with some elementary functions (for instance, the successor function $n(x) = x + 1$) which are obviously computable. All these definitions of computability by TMs, lambda calculus, recursive functions, etc. can be proved to be mathematically equivalent. Obviously, each of these precise concepts defines a procedure which is intuitively effective.

Alonzo Church thus postulated his famous thesis that the informal intuitive notion of an effective procedure is identical with one of these equivalent precise concepts, such as that of a TM (Church 1936). With respect to the definition of a UTM, the *Church–Turing thesis* can also be formulated in the following way:

> The class of all intuitively computable mathematical functions coincides with the class of all mathematical functions computable on a UTM.

Of course, this statement cannot be proved, because mathematically precise concepts are being compared with an informal intuitive notion. Nevertheless, the mathematical equivalence of several precise *concepts of computability* which are intuitively effective, confirms the Church–Turing thesis. Consequently, we can speak about computability, effectiveness, and computable functions without referring to particular effective procedures ("algorithms") such as TMs, recursive functions, etc. According to the Church–Turing thesis, we may, in particular, say that every computational procedure (algorithm) may be calculated by a TM. So every recursive function, as a kind of machine program, may be calculated by a general purpose computer.

Now we are able to define effective procedures of decision and enumerability, which were already demanded by Leibniz' program of a *mathesis universalis*. The characteristic function f_M of a subset M of natural numbers is defined as $f_M(x) = 1$ if x is an element of M, and as $f_M(x) = 0$ otherwise. Thus, a set M is defined as *effectively decidable* if its characteristic function saying whether or not a number belongs to M is effectively computable (or recursive).

A set M is defined as *effectively (recursively) enumerable* if there exists an effective (recursive) procedure f for generating its elements, one after another (formally $f(1) = x_1, f(2) = x_2, \dots$ for all elements $x_1, x_2, \dots$ from M). It can easily be proved that every recursive (decidable) set is recursively enumerable. But there are recursively enumerable sets which are not decidable. These are the first hints that there are limits to Leibniz' original optimistic program, based on the belief in universal decision procedures.

In his famous paper "On Computable Numbers, with an Application to the *Entscheidungsproblem*", Turing reformulated Kurt Gödel's 1931 (Feferman 1986) results on the *limits of proof and computation*, replacing Gödel's universal arithmetic-based formal language with the formal and simple devices of a TM (Turing 1936/1937). He argued that a TM would be capable of performing any conceivable mathematical computation if it were representable as an algorithm. A fundamental insight was his proof that there was no solution to the *Entscheidungsproblem* by first showing that the *halting problem* for TMs is *undecidable*: it is not possible to decide, in general, algorithmically whether a given TM will ever halt. While his proof was published subsequent to Alonzo Church's equivalent proof in respect to his lambda calculus, Turing was unaware of Church's work at the time.

From September 1936 to July 1938 he spent most of his time at the Institute for Advanced Study in Princeton studying under Alonzo Church. In June 1938 he obtained his Ph.D. from Princeton. His dissertation introduced the notion of relative computing, where TMs are augmented with so-called oracles, allowing a study of problems that cannot be solved by a TM (Turing 1939, Sect. 4). This concept will be important for the final question of our book: Is the Universe a computer?

There is another of Alan Turing's concepts that is central to our book. From 1952 until his death in 1954, Turing worked on mathematical biology, specifically morphogenesis. He published one paper on the subject called "The Chemical Basis of Morphogenesis" in 1952, putting forth the *Turing hypothesis of pattern formation* (Turing 1952). His central interest in the field was to understand Fibonacci phyllotaxis, the existence of Fibonacci numbers in plant structures. He used reaction–diffusion equations to model pattern formation. Later papers went unpublished until 1992 when "Collected Works of A.M. Turing" was published (Gandy et al. 1992–2001).

The Birth of Cellular Automata

Pattern formation was in the air of different research fields. In the 1940, Stanisław Ulam, while working at the Los Alamos National Laboratory, studied the growth of crystals, using a simple lattice network as his model. At the same time, John von Neumann, Ulam's colleague at Los Alamos, was working on the problem of self-replicating systems. Von Neumann's initial design was founded upon the notion of one robot building another robot. But John von Neumann had to realize the great technical difficulties of building a self-replicating robot. Ulam suggested that von Neumann should develop his design as a mathematical abstraction, such as the one Ulam used to study crystal growth. Ulam's suggestion led to the first system of cellular automata (Burks 1970). Like his lattice network of crystals, von Neumann's *cellular automata* are two-dimensional (Neumann 1966), with his *self-replicator* implemented algorithmically. The result was a universal copier and constructor working within a cellular automaton (CA) with a small neighborhood, and with 29 states per cell. Von Neumann gave an existence proof that a particular pattern would make endless copies of itself within the given cellular universe. This design is known as the tessellation model, and is called a *von Neumann universal constructor*. Also in the 1940s, Norbert Wiener and Arturo Rosenblueth developed a CA model of excitable media. Their specific motivation was the mathematical description of impulse conduction in cardiac systems (Wiener et al. 1946).

In general, a *CA* can be imagined as a chessboard-like grid of cells changing their states (indicated, for example, by different colors or numbers) with respect to simple rules depending on the states of neighboring cells. In nature and technology, alternative cellular states may be realized, such as by spontaneously produced and annihilated elementary particles; switched-on and switched-off genes; living and dead cells of organisms; firing and non-firing neurons of brains; or the electrical switches of circuits. Their interaction leads to the emergence of complex patterns that grow, die, and sometimes even reproduce themselves like cellular organisms. They follow a *local principle of cellular activity* in a certain environment.

In the 1970s a two-state, two-dimensional CA named the *Game of Life* became very widely known. Invented by John Conway and popularized by Martin Gardner in a *Scientific American* article (Gardner 1970, 1971), its rules are as follows: If a cell has two black neighbors, it stays the same. If it has three black neighbors, it remains black if previously black, but changes to black if previously white. In all other situations it becomes white. Despite its simplicity, which can be understood by school-age boys and girls, the system achieves an impressive complexity of behavior, fluctuating between apparent randomness and order. The expanding complexity of new patterns reminds us of the Cambrian explosion of new organisms undergoing Darwinian evolution. One of the most apparent features of the Game of Life is the frequent occurrence of gliders, arrangements of cells that essentially move themselves across the grid. It is possible to arrange the automaton so that the gliders interact to perform computations. It was a breakthrough when it was proved that the Game of Life is a UTM. This proof seemed to suggest that biological evolution itself could be a universal CA.

The Zuse-Fredkin Thesis

In 1967, the German computer pioneer Konrad Zuse (1910–1995) suggested that the entire universe could be computed on a computer in the kind of a CA. Historically, it is remarkable that he also built the first programmable computers (1935–1941) and devised the first higher-level programming language (1945). In 1969, Konrad Zuse published his book *Calculating Space*, proposing that the physical laws of the universe are discrete, and that the entire universe is the output of a deterministic computation on a giant CA. This was the first book on what today is called digital physics. "Calculating Space" is the title of MIT's English translation of the German book *Rechnender Raum* (literally: "space that is computing") (Zuse 1969). He focused on cellular automata as a possible substrate of the computation.

In 1967 Zuse carefully remarked (p. 337) that at the moment there were no full *digital models of physics*, but that did not prevent him from asking for the consequences of a total discretization of all natural laws (Schmidhuber 1997). For lack of a complete automata-theoretic description of the universe he tried to study several simplified models. Similar to von Neumann's approach, he considered neighboring cells that update their values based on surrounding cells, implementing the spread and creation and annihilation of elementary particles. He mentioned (p. 341) that the term "CA" had already been introduced in the literature at that time and cited von Neumann's 1966 book "Theory of Self-reproducing Automata" (Neumann 1966). In Chap. 4, Konrad Zuse studied the relation of cellular automata to relativity theory, information theory, probability theory, and the concepts of determinism and causality. Concerning information, he strongly believed that in the cosmos as a big computer and a closed system, the information content cannot increase.

After Ulam's and von Neumann's concept of cellular automata, the idea of *digital physics* seemed to be in the air during the 1960s. Independently of Zuse, Edward Fredkin (born 1934), a US Air Force jet-fighter pilot, programmer (PDP-1 assembler), professor of physics and CEO of a diverse set of companies, also suggested that the Universe itself is some kind of computational device, namely, a highly parallel computational machine like a *CA*. Fredkin's conjecture remained unpublished until the appearance of John Conway's "Game of Life" in 1969, and subsequent popular publications by Martin Gardner on the Game of Life and cellular automata in general. It was Gardner (1970, 1971) who first made popular the "Universe as CA" idea in his February 1971 article in the *Scientific American* series, which later appeared in book form. The first publication dedicated to Fredkin's thesis appeared only in 1988 in form of an interview given to Robert Wright, which was later enlarged and published in a book (Wright 1989). In 1990 Fredkin himself wrote a paper on his thesis in the respected scientific journal *Physica* (Fredkin 1990). Nevertheless, he also advocates a kind of *digital philosophy* (*see* "Dr. Edward Fredkin's Digital Philosophy site": http://www.digitalphilosophy.org). Obviously, like Zuse, he is a digital practitioner with philosophical visions. With respect to Zuse's contribution, it is fair to use the description "*Zuse-Fredkin thesis*", which proclaims:

The Universe is a cellular automaton.

Cellular automata may be proven to be equivalent to TMs. There are universal cellular automata simulating all kinds of cellular automata. Obviously, the *Zuse-Fredkin thesis* includes the *Church–Turing thesis* (Petrov 2003). If the universe is a CA (in the sense of the Zuse-Fredkin thesis), then everything in the universe is a computational process simulated by a universal CA. Humans with their mathematical thoughts would be only special systems in the universe, and realized by some kind of algorithms. Therefore, intuitively computable functions (such as the mathematical thoughts of humans) can be simulated by a UTM, in the sense of the Church–Turing thesis. But, the reverse conclusion is more problematic. This conclusion depends on the open questions that: (a) all processes in the universe can actually be represented by mathematical laws and theories and, (b) all mathematical laws and theories of the universe can be simulated by computational algorithms in the sense of the Church–Turing thesis.

A New Kind of Science?

In 1983 Stephen Wolfram published the first of a series of papers systematically investigating a very basic but essentially unknown class of cellular automata, which he terms elementary cellular automata (Wolfram 1986). The unexpected *complexity of the pattern formation* according to these simple rules led Wolfram to suspect that complexity in nature may be caused by similar mechanisms (Wolfram 1994). In addition, during this period Wolfram speculated about the concepts of intrinsic randomness and computational irreducibility. He suggested that a special elementary CA (rule 110) might be *universal.* That was proved later by Wolfram's research assistant Matthew Cook in the 1990s.

In 2002 Wolfram published the 1280-page text book *A New Kind of Science*, which extensively argues that the new discoveries about cellular automata depend on *computational experiments* with high speed computers, not on mathematical proofs (Wolfram 2002). Therefore, his book seemed to proclaim a new kind of future *methodology*, namely experiments with computers instead of "traditional" mathematical equations. The computational experimental approach would prevail over the classical *analytical style of science*. But the lack of analytical explanation and confirmation was perhaps one essential point why Wolfram's book was not widely distributed in the physical sciences, although it presented an impressive variety of pattern formations. From a methodological point of view, it is true that modern science is mainly made possible by computational data mining and computational modeling, because of the increasing complexity of research fields. But that does not mean that we can abandon mathematical explanations and confirmation by analytical proofs and the solution of equations.

At this point, our book comes in. We also consider elementary (1-dimensional) cellular automata (Chap. 2). We are also impressed by the complex variety of pattern formation, with the emergence of sometimes surprising new structures. But we do not only describe and classify cellular patterns qualitatively from a *phenomenological point of view*. This would be *Aristotelian science* before Galileo, systematizing and categorizing plants and animals with respect to their observable common properties. To understand and explain the emergence of phenomena, we must analyze the laws of pattern formation represented by mathematical equations. The laws are confirmed in computer experiments. This kind of *analytical methodology* is similar to mathematical physics, although the models are digital. A nice example is the *complexity index* of cellular automata, which allows a precise classification of (1-dimensional) cellular automata (Chap. 3). In addition, the equations reveal precise *mathematical symmetries* in the universe of cellular automata (Chap. 4). Symmetries lead to *equivalence classes,* which mean an immense reduction of research complexity. In these cases, we can restrict our investigations to representative examples of equivalence classes. There is one distinguished symmetry class of cellular automata with universal computability. Wolfram's CA 110 is only one of four examples. With respect to its universal capacities and symmetries, this equivalence class is considered the *holy grail* in the universe of (1-dimensional) cellular automata.

The fundamental question arises how far *digital phenomena* can be mapped on to the *continuous dynamics of complex systems.* This is broadly possible (apart from some aspects which are explicitly discussed in our book). Therefore, the whole impact of the highly developed mathematical theory of complex dynamical systems can be used to find precise and exact classifications and predictions in the universe of cellular automata (Chap. 5). Remarkable results concern the problem of reversible and irreversible time (*time arrow*) in the universe of cellular automata, which can be decided in a mathematically precise way (Chap. 6). In general, cellular automata are a powerful instrument for modeling all aspects of complex dynamics (Toffoli 1987; Hoekstra et al. 2010). Despite the simple local rules for cellular activities, highly sophisticated mathematics allows the analysis of the *nonlinear dynamics of complex systems.*

In any case, the universe of elementary (1-dimensional) cellular automata is a nice field for exercises in *digital physics* aiming to understand and illustrate fundamental problems of the physical world through computational modeling. Digital physics is an important field of research, even without the somewhat metaphysical belief in the *Zuse-Fredkin thesis.* A typical example is the explanation of *emergence* and *self-organization* of complex patterns with the dynamics of cellular automata. Very simple rules of behavior can lead to highly complex structures, which cannot be forecast in the long run. This is not only a hypothesis on the basis of observations, but a mathematically precise proof in the case of cellular automata with universal computability. With this background, it becomes understandable that we need no *intelligent design* of complex structures, but only very simple rules for local elements that generate global complex structures during their evolution.

Complex Dynamical Systems

We strongly believe that twenty first century science is a *science of complexity* (Chua 1998; Mainzer 2005, 2007a, b, 2009). The challenges of complexity research are already mirrored in cellular automata. We therefore introduce some basic concepts of complex dynamical systems, which are defined precisely later in the context of cellular automata. A *dynamical system* is characterized by its elements and the time-dependent development of their states. A dynamical system is called *complex* if many (more than two) elements interact in causal feedback loops generating unstable states, chaos or other kinds of attractors. The *states* may refer to moving planets, molecules in a gas, gene expressions of proteins in cells, excitation of neurons in a neural net, nutrition of populations in an ecological system, or products in a market system. The *dynamics* of a system, i.e., the change of a system's states depending on time, is mathematically described by *differential equations*. A *conservative* (Hamiltonian) system, such as an ideal pendulum, is determined by the reversibility of the direction of time and the conservation of energy. *Dissipative* systems, a real pendulum with friction, for example, are irreversible.

In classical physics, the dynamics of a system is considered a continuous process. But continuity is only a mathematical idealization. Actually, a scientist has single observations or measurements at discrete points in time, which are chosen to be equidistant or defined by other measurement devices. In discrete processes, there are finite differences between the measured states and none of the infinitely small differences (differentials) that are assumed in a continuous process. Thus, discrete processes are mathematically described by *difference equations*.

Random events (such as Brownian motion in a fluid, mutation in evolution, innovations in an economy) are represented by additional fluctuation terms. *Classical stochastic processes*, for example, the billions of unknown molecular states in a fluid, are defined by time-dependent differential equations with distribution functions for probabilistic states. In *quantum systems* of elementary particles, the dynamics of quantum states is defined by Schrödinger's equation with observables (such as position and momentum of a particle) depending on Heisenberg's uncertainty principle, which allows only probabilistic forecasts of future states.

Historically, during the centuries of classical physics, the universe was considered a deterministic and conservative system. The astronomer and mathematician Laplace, for example, assumed the total computability and predictability of nature if all natural laws and initial states of celestial bodies were known. The *Laplacean spirit* expressed the belief of philosophers in determinism and computability of the world during the eighteenth and nineteenth centuries.

Laplace was right about *linear* and *conservative dynamical systems*. In general, a linear relation means that the rate of change in a system is proportional to its cause: Small changes cause small effects, while large changes cause large effects. Changes of a dynamical system may be modeled in one dimension by changing values of a time-dependent quantity along the time axis (*time series*). Mathematically, linear equations are completely computable. This is the deeper reason for Laplace's philosophical assumption being right for linear and conservative systems.

In systems theory, the complete information about a dynamical system at a certain time is determined by its *state* at that time. If the state of a dynamical system is determined by more than two quantities, a higher dimensional phase space is needed to study the dynamics of the system. From a methodological point of view, time series and phase spaces are important instruments for studying system dynamics. The state space of a system contains the complete information of its past, present and future behavior.

At the end of the nineteenth century, Poincaré discovered that celestial mechanics is not a completely computable clockwork, even if it is considered as a deterministic and conservative system (Poincaré 1897). The mutual gravitational interactions of more than two celestial bodies (*the many-body-problem*) may be represented by causal feedback loops corresponding to nonlinear and non-integrable equations with instabilities and irregularities. In a strictly dynamical sense, the degree of complexity depends on the degree of *nonlinearity* of a dynamical system. According to the Laplacean view, similar causes effectively determine similar effects. Thus, in the phase space, trajectories that start close to each other also remain close to each other during their evolution with time. Dynamical systems with deterministic chaos exhibit an exponential dependence on initial conditions for bounded orbits: The separation of trajectories with close initial states increases exponentially.

Thus, tiny deviations in initial data lead to an exponential increase in computational efforts for future data, limiting long-term predictions, although, in principle, the dynamics are uniquely determined. This is known as the '*butterfly effect*': Initial, small, and local causes soon lead to unpredictable, large, and global effects. According to the famous *KAM-Theorem* of Kolmogorov (1954); Arnold (1963), and Moser (1967), trajectories in the phase space of classical mechanics are neither completely regular, nor completely irregular, but depend sensitively on the chosen initial conditions (Arnold 1963; Kolmogorov 1954; Moser 1967).

Dynamical systems may be classified on the basis of the effects of the dynamics on a region of the *phase space* (Alligood et al. 1996). A *conservative system* is defined by the fact that, during evolution with time, the volume of a region remains constant, although its shape may be transformed. In a *dissipative* system, dynamics causes a contraction in volume.

An *attractor* is a region of a phase space into which all trajectories departing from an adjacent region (the so-called basin of attraction), tend to converge. There are different kinds of attractors. The simplest class of attractors contains *fixed points*. In this case, all trajectories of adjacent regions converge to a point. An example is a dissipative harmonic oscillator with friction: The oscillating system is gradually slowed down by frictional forces and finally come to a rest at an equilibrium point.

A second class contains periodic attractors with non-empty basins of attraction, which are called limit cycles. In phase space, a limit-cycle is a closed trajectory having the property that at least one other trajectory spirals into it either as time approaches infinity or as time approaches negative infinity. In the case where all the neighboring trajectories approach the limit cycle as time approaches infinity, it is called a stable or attractive limit cycle (ω-limit cycle). If, instead, all neighboring trajectories approach it as time approaches negative infinity, it is an unstable or non-attractive limit-cycle (α-limit cycle). Stable limit-cycles imply self-sustained oscillations. Any small perturbation from the closed trajectory would cause the system to return to the limit-cycle, making the system stick to the limit-cycle. An example of a stable limit cycle is a Van der Pol oscillator modeling a simple vacuum-tube oscillator circuit. For a simple dynamical system with only two degrees of freedom and continuous time, the only possible attractors are fixed points or periodic limit cycles.

In continuous systems with a phase space of dimension $n > 2$, more complex attractors are possible. Dynamical systems with quasi-periodic limit cycles show a time evolution which can be decomposed into different periodic parts without a unique periodic regime. The corresponding time series consist of periodic parts of oscillation without a common structure. Nevertheless, closely starting trajectories remain close to each other during time evolution. The third class contains dynamical systems with *chaotic attractors* which are non-periodic, with an exponential dependence on initial conditions for bounded orbits. A famous example is the chaotic attractor of a Lorenz system with some qualitative similarity to the chaotic development of weather caused by local events, which cannot be forecast in the long run (the butterfly effect).

Measurements are often contaminated by unwanted noise, which must be separated from the signals of specific interest. In addition, to forecast the behavior of a system, the development of its future states must be reconstructed in a corresponding phase space from a finite sequence of measurements. Thus, time-series analysis is an immense challenge in different fields of research ranging from fields such as climatic data in meteorology, ECG-signals in cardiology, and EEG-data in brain research, to economic data in economics and finance (Abarbanel 1995; Small 2005). Beyond the patterns of dynamical attractors, randomness of data must be classified by statistical distribution functions.

Typical phenomena found in our world, such as weather, climate, the economy and daily life are much too complex for a simple deterministic description to exist. Even if there is no doubt about the deterministic evolution of, the atmosphere, for example, the current state, knowledge of which would be needed for a deterministic prediction, contains too many variables to be measurable with sufficient accuracy. Hence, our knowledge does not usually suffice for a deterministic model. Instead, very often a stochastic approach is more appropriate. Ignoring the unobservable details of a complex system, we accept a lack of knowledge. Depending on the unobserved details, the observable part may evolve in different ways. However, if we assume a given probability distribution for the unobserved details, then the different evolutionary paths of the observables also emerge with specific, associated probabilities. Thus, the lack of knowledge about the system prevents us from deterministic predictions, but does allow us to assign probabilities to the different possible future states. It is the task of a time series analysis to extract the necessary information from past data.

Complex models contain nonlinear feedback, and the solutions to these are usually obtained by numerical methods (Bungartz et al. 2009). Complex statistical models are data driven and try to fit a given set of data using various distribution functions. There are also hybrids, coupling dynamic and statistical aspects, including deterministic and stochastic elements. Simulations are often based on computer programs, connecting input and output in nonlinear ways. In this case, models are calibrated by training the programs, in order to minimize the error between the output and the given test data.

In the simplest case of statistical distribution functions, a Gaussian distribution has exponential tails situated symmetrically to the far left and far right of the peak value. Extreme events (such as disasters, pandemics, or floods) occur in the tails of the probability distributions (Albeverio et al. 2006). Contrary to the Gaussian distribution, probabilistic functions $p(x)$ of heavy tails with extreme fluctuations are mathematically characterized by power laws, e.g., $p(x) \sim x^{-\alpha}$ with $\alpha > 0$. Power laws possess scale invariance corresponding to the (at least statistical) self-similarity of their time series of data. Mathematically, this property can be expressed as $p(bx) = b^{-\alpha}p(x)$ meaning that the change of variable x to bx results in a scaling factor independent of x while the shape of distribution p is conserved. So, power laws represent scale-free complex systems. The Gutenberg–Richter size distribution of earthquakes is a typical example of natural sciences. Historically, Pareto's distribution law of wealth was the first power law in the social sciences, with a fraction of people presumably several times wealthier than the mass population of a nation (Mainzer 2007a, b, 2008).

Beyond the Church–Turing Thesis?

Why should we restrict ourselves to cellular automata in modeling complex dynamical systems? Are there any types of computation that cannot be done by a TM, but that can be done using some other kind of physical or abstract machine? These types of machine would realize so-called hyper-computation. These models of computation are capable of solving problems (e.g., the halting problem) that cannot be solved using TMs or cellular automata.

It is historically remarkable that Turing himself took the first steps beyond the Church–Turing thesis. In his dissertation, he introduced the oracle TMs (Turing 1939). These are equipped with an oracle to answer questions, such as to the halting problem. In general, their computational procedures depend on the application of a function the algorithm of which is not known, or in other words, an "oracle". Thus, computing a function f by an oracle TM means that the computation of f is realized by a TM depending on a certain additional device (or function) g the algorithm of which is not known. Therefore, f is also called computable relative to a function ("oracle") g. Turing's concept of relative computability does not violate the Church–Turing thesis, but enlarges it. Therefore, one can consider a relative Church–Turing thesis (Mainzer 1973) which could, of course, also be applied to oracle cellular automata.

The concept of relative computability is less speculative than it seems to be at a first glance. Although, from a technical point of view, we have no idea how to construct an "oracle", there are useful applications of modeling with relative computability. For example, in all kinds of everyday scenarios we may decide without full information of all the details. Then, our decision depends on unknown "oracles". In economy, Herbert S. Simon, Nobel prize winner for economics and one of the fathers of Artificial Intelligence, introduced the term "bounded rationality" to describe these realistic situations. In mathematics, the existence of functions depends on the axioms of mathematical theories. An important example is the axiom of choice, which only guarantees the existence of a function, but not an effective procedure to compute it. In this case, a mathematical proof is only relatively computable. In general, mathematical proofs consist of computable (constructive) and non-computable (non-constructive) parts. Thus, relative computability is a useful concept to describe mathematical praxis.

What about situations when answers cannot be known in principle, for example, in the case of a fair coin toss? A probabilistic TM is like a nondeterministic TM, except that it chooses its next transition uniformly at random from the legal choices (Wolfram 2002). Theoretically, a probabilistic TM can generate truly random bits, a capability not possible on a conventional TM. From a physical point of view, possible candidates of probabilistic TMs producing randomness are quantum processes (for example, β-decay). Supporters of the Zuse-Fredkin thesis would argue that nature only generates pseudo-random numbers and these could be simulated on a deterministic TM.

But, Richard Feynman's arguments should be taken seriously in that the problem of simulating quantum physics with a computer based on classical physics appears to be intractable, thereby suggesting that quantum computers may be inherently more powerful than classical computers. It is historically remarkable that he published his article "Simulating Physics with Computers" in 1982 (Feynman 1982), when Aspect's experiments confirmed Bell's inequalities (Audretsch et al. 1996). Thus, there is a difference between classical probabilities and quantum probabilities: Quantum systems cannot be simulated by classical probabilistic TMs. In the meantime, quantum computing has been logically and mathematically well developed.

In Chap. 7, we therefore introduce basic concepts of quantum physics, and analyze the concept of (1-dimensional) quantum cellular automata that may be proven to be equivalent to quantum circuits and quantum TMs. There is also a universal quantum CA or quantum TM. In general, the Church–Turing thesis, together with quantum and classical complexity theory, will be discussed in relation to quantum physics.

In the age of computerization, digital models are applied in nearly all scientific disciplines. After physics, we will consider modeling with cellular automata in the life sciences (Chap. 8) and neurosciences (Chap. 9). Systems biology is a recent addition to systems science, analyzing molecular and cellular systems of increasing complexity (Mainzer 2010). Cellular automata deliver models of complex networks of genes and proteins (for example), to analyze attractor dynamics within cells. Cellular automata can also be considered as neural networks of firing and non-firing neurons, generating cell assemblies that are correlated with mental or cognitive states. Thus, in the end, it is not essential whether the universe is, or is not, an automaton in any metaphysical sense. It is impressive how far the universe may be mapped by digital models that are made possible by human technology's high-speed computers. The concept of cellular automata historically initiated the development of digital models and is still a mirror of key problems in our universe.

References

H.D.I. Abarbanel, *Analysis of Observed Chaotic Data* (Springer, New York, 1995)

S. Albeverio, V. Jentsch, H. Kantz (eds.), *Extreme Events in Nature and Society* (Springer, Berlin, 2006)

K.T. Alligood, T.D. Sauer, J.A. Yorke, *Chaos: An Introduction to Dynamical Systems* (Springer, New York, 1996)

V.I. Arnold, Small denominators II, proof of a theorem of A.N. Kolmogorov on the preservation of conditionally-periodic motions under a small perturbation of the Hamiltonian. Russ. Math. Surveys **18**, 5 (1963)

J. Audretsch, K. Mainzer (Hrsg.) *Wieviele Leben hat Schrödingers Katze? Zur Physik und Philosophie der Quantenmechanik,* 2nd edn. (Spektrum Akademischer Verlag, Heidelberg, 1996)

H.-J. Bungartz, S. Zimmer, M. Buchholz, D. Pflüger, *Modellbildung und Simulation. Eine anwendungsorientierte Einführung* (Springer, Berlin, 2009)

A.W. Burks (ed.), *Cellular Automata* (University of Illinois Press, Urbana, 1970)

L.O. Chua, *CNN: A Paradigm for Complexity* (World Scientific, Singapore, 1998)

A. Church, A note on the Entscheidungsproblem. J. Symb. Log. **1**, 40–41 (1936)

S. Feferman (ed.), *Kurt Gödel: Collected Works* (Oxford University Press, Oxford, 1986), pp. 144–195

R. Feynman, Simulating physics with computers. Int. J. Theor. Phys. **21**(6–7), 467–488 (1982)

E. Fredkin, Digital Mechanics: An informational process based on reversible universal CA. Physica D 45:254–270 (1990)

R. Gandy, C.E.M. Yates (eds.) *Collected Works of A. M. Turing* (North-Holland, Amsterdam 1992–2001)

M. Gardner, The fantastic combinations of John Conway's new solitaire game of life. Sci. Am. **223**, 120–123 (1970)

M. Gardner, Mathematical games: on cellular automata, self-reproduction, the Garden of Eden, and the game "Life". Sci. Am. **224**, 112–117 (1971)

R. Herken (ed.) *The Universal Turing Machine. A Half-Century Survey,* 2nd edn. (Springer, Wien 1995)

A.G. Hoekstra, J. Kroc, P.M.A. Sloot (eds.), *Simulating Complex Systems by Cellular Automata* (Springer, Berlin, 2010)

A.N. Kolmogorov, On conservation of conditionally-periodic motions for a small change in Hamilton's function. Dokl. Akad. Nauk. USSR **98**, 525 (1954)

K. Mainzer, *Computer—Neue Flügel des Geistes?* (De Gruyter, Berlin 1994)

K. Mainzer, *Computerphilosophie* (Junius, Hamburg, 2003)

K. Mainzer, *Symmetry and Complexity: The Spirit and Beauty of Nonlinear Science* (World Scientific, Singapore, 2005)

K. Mainzer, *Thinking in Complexity. The Computational Dynamics of Matter, Mind, and Mankind,* 5th edn. (Springer, Berlin, 2007a)

K. Mainzer, *Der kreative Zufall. Wie das Neue in die Welt kommt* (C.H. Beck, München, 2007b)

K. Mainzer, *Komplexität* (UTB-Profile, Paderborn, 2008)

K. Mainzer (ed.), Complexity. Eur. Rev. (Academia Europaea) **17**(2), 219–452 (2009)

K. Mainzer, *Leben als Maschine? Von der Systembiologie zur Robotik und Künstlichen Intelligenz* (Mentis, Paderborn, 2010)

K. Mainzer, *Mathematischer Konstruktivismus.* Dissertation Universität Münster (1973)

M. Minsky, Size and structure of universal Turing machines using tag systems, recursive function theory. Proceedings Symposium in Pure Mathematics, vol. 5 (American Mathematical Society, Providence RI), pp. 229–238 (1962)

J. Moser, Convergent series expansions of quasi-periodic motions. Math. Ann. **169**, 163 (1967)

J.V. Neumann, *Theory of Self-Reproducing Automata* (University of Illinois Press, Urbana, 1966)

P. Petrov, Church-Turing thesis as an immature form of Zuse-Fredkin thesis. (More arguments in favour of the 'universe as a cellular automaton' idea). 3rd WSEAS *International Conference on Systems Theory and Scientific Computation*. Special session on cellular automata and applications (2003) (online version: http://digitalphysics.org/Publications/Petrov/Pet02a2/Pet02a2.htm)

H. Poincaré, *Les Méthodes Nouvelles de la Méchanique Céleste I-II* (Gauther-Villars, Paris, 1897)

J. Schmidhuber, A computer scientist's view of life, the universe, and everything. Lecture Notes in Computer Science (Springer, Berlin, 1997), pp. 201–208

C. Shannon, in *A Universal Turing Machine With Two Internal States*. Automata Studies (Princeton University Press, Princeton, 1956), pp. 157–165

M. Small, *Applied Nonlinear Time Series Analysis: Applications in Physics, Physiology and Finance* (World Scientific, Singapore, 2005)

T. Toffoli, N. Margolus, *Cellular Automata Machines: A New Environment for Modeling* (MIT Press, Cambridge, 1987)

B.A. Toole, *Ada, the Enchantress of Numbers* (Strawberry Press, Mill Valley, CA, 1992)

A.M. Turing, Computability and λ-definability. J. Symb. Log. **2**, 153–163 (1937)

A.M. Turing, Systems of logic based on ordinals. Proc. Lond. Math. Soc. **2**(45), 161–228 (1939)

A.M. Turing, The chemical basis of morphogenesis. Philos. Trans. Royal Soc. Lond. Ser. B, Biol. Sci. **237**(641), 37–72 (1952)

A.M. Turing, On computable numbers with an application to the Entscheidungsproblem. Proc. Lond. Math. Soc. **42**(2) 230–265, corrections, ibid, **43**, 544–546 (1936/1937)

N. Wiener, A. Rosenbluth, The mathematical formulation of the problem of conduction of impulses in a network of connected excitable elements, specifically in cardiac muscle. Arch. Inst. Cardiol. Mex. **16**, 205–265 (1946)

S. Wolfram, *Theory and Applications of Cellular Automata* (World Scientific, Singapore, 1986)

S. Wolfram, *Cellular Automata and Complexity* (Addison–Wesley, Reading, 1994)

S. Wolfram, *A New Kind of Science* (Wolfram Media, Champaign, 2002)

R. Wright, *Three Scientists and Their Gods: Looking for Meaning in an Age of Information* (HarperCollins, New York, 1989)

K. Zuse, *Rechnender Raum* (Friedrich Vieweg & Sohn, Braunschweig, 1969)

Chapter 2
Simplicity in the Universe of Cellular Automata

Because of their simplicity, rules of cellular automata can easily be understood. In a very simple version, we consider *two-state one-dimensional* (1D) *cellular automata* (CA) made of identical cells with a *periodic boundary condition*. In this case, the object of study is a ring of coupled cells with $L = I + 1$ cells, labeled consecutively from $i = 0$ to $i = I$ (Fig. 1a). Each cell i has two *states* $u_i \in \{0, 1\}$, which are coded by the colors blue and red, respectively. A clock sets the pace in discrete intervals by iterations or generations. The state u_i^{t+1} of all i at time $t + 1$ (i.e. the next generation) is determined by the states of its nearest neighbors u_{i-1}^t, u_{i+1}^t, and itself u_i^t at time t (Fig. 1c), that is, by a *Boolean function* $u_i^{t+1} = N(u_{i-1}^t, u_i^t, u_{i+1}^t)$, in accordance with a prescribed *Boolean truth table* of $8 = 2^3$ distinct 3-input patterns (Fig. 1d).

From Simple Local Rules to Global Complex Patterns

These eight 3-input patterns can nicely be mapped into the eight vertices of a toy cube (Fig. 1b), henceforth called a *Boolean cube* (Chua et al. 2002). The output of each prescribed 3-input pattern is mapped onto the corresponding colors (red for 1, blue for 0) at the vertices of the Boolean cube (in Fig. 1d yet unspecified). Because there are $2^8 = 256$ distinct combinations of 8 bits, there are exactly 256 Boolean cubes with distinct vertex color combinations. Thus we get a gallery of picturesque toy cubes.

It is convenient to associate the 8-bit patterns of each Boolean function with a decimal number N representing the corresponding 8-bit word, namely $N = \beta_7 \cdot 2^7 + \beta_6 \cdot 2^6 + \beta_5 \cdot 2^5 + \beta_4 \cdot 2^4 + \beta_3 \cdot 2^3 + \beta_2 \cdot 2^2 + \beta_1 \cdot 2^1 + \beta_0 \cdot 2^0$ with $\beta_i \in \{0, 1\}$. Notice that since $\beta_i = 0$ for each blue vertex in Fig. 1b, N is simply obtained by adding the weights (indicated next to each pattern in Fig. 1b) associated with all red vertices. For example, for the Boolean cube shown in Fig. 2b,

K. Mainzer and L. Chua, *The Universe as Automaton*, SpringerBriefs in Complexity, DOI: 10.1007/978-3-642-23477-4_2, © The Author(s) 2012

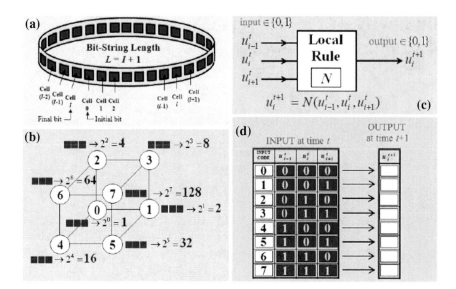

Fig. 1 Scheme of a two-state one-dimensional cellular automaton with local rule N

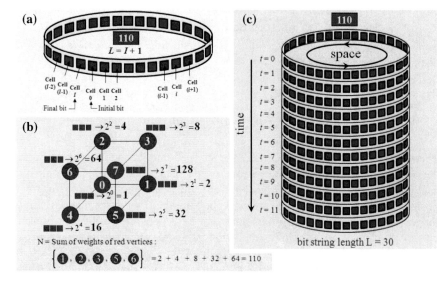

Fig. 2 Example of local rule 110

we have $N = 0 \cdot 2^7 + 1 \cdot 2^6 + 1 \cdot 2^5 + 0 \cdot 2^4 + 1 \cdot 2^3 + 1 \cdot 2^2 + 1 \cdot 2^1 + 0 \cdot 2^0 = 2^6 + 2^5 + 2^3 + 2^2 + 2^1 = 110$.

For the example of local rule 110, the ring and the colored vertices of the Boolean cube are shown in Fig. 2a, b. Given any initial binary bit-configuration at time $t = 0$, the local rule N is used to update the state of each cell i at time $t + 1$, using the states of the three neighboring cells $i - 1$, i, and $i + 1$, centered at location i, respectively. The *space–time pattern* for the initial state is shown in Fig. 2c for $t = 0, 1, 2, \ldots, 11$.

In principle, one can draw and paint the patterns of CA following these rules step by step. Modern computers with high speed and capacity allow extensive computer experiments to study pattern formations of these automata. Stephen Wolfram discovered remarkable analogies with patterns in physics and biology (Wolfram 2002). In the world of CA many phenomena of the physical world seem to evolve. Some automata generate symmetric patterns reminding us of the coloring in sea shells, skins or feathers. Other automata reproduce rhythms like oscillating waves. Some of these automata stop their development after a finite number of steps, independent of their initial state, and remain in a constant color state like a system reaching an equilibrium state for all future steps. Some automata develop complex patterns reminding us of the growth of corals or plants, depending sensitively on tiny changes to the initial states. This phenomenon is well-known as the butterfly-effect, when local events lead to global effects in chaotic and unstable situations (such as weather and climate). Even these chaotic patterns can be generated by CA.

One can try to classify these patterns with respect to their outward appearance just as zoologists and botanists distinguish birds and plants in taxonomies. This observational method is used by Wolfram for patterns experimentally generated by computers. But sometimes outward features are misleading. The fundamental question arises: Are there laws of complex pattern formation for CA like those in nature? Can the development of complex patterns be predicted in a mathematically rigorous way as in physics? We argue for a mathematically precise explanation of the dynamics in CA. Therefore, they must also be characterized by complex dynamical systems determined with differential equations like those in physics. This is, of course, beyond the scope of elementary rules of toy worlds. But we should keep this perspective in mind.

Cellular Automata as Dynamical Systems

For maximum generality, each cell i is assumed to be a *dynamical system* with an intrinsic state x_i, an output y_i, and *three inputs* u_{i-1}, u_i, and u_{i+1} where u_{i-1} denotes the input coming from the left neighboring cell $i - 1$, u_i denotes the self input of cell i, and $i + 1$ denotes the input coming from the right neighboring cell $i + 1$ in the ring of Fig. 1a. Each cell evolves with its prescribed dynamics and its own time scale. When coupled together, the system evolves consistently with its own rule as well as the rule of interaction imposed by the coupling laws.

Each *input* is assumed to be a *constant integer* $u_i \in \{-1, 1\}$, and the *output* y_i *converges* to a *constant* either -1 or 1 from a zero initial condition $x_i(0) = 0$. Actually, it takes a finite amount of time for any dynamical system to converge to an *attractor*. But, for the purpose of idealized CA, each attractor is assumed to be reached instantaneously. Under this assumption and with respect to the *binary input* and *output*, our *dynamical system* can be defined by a *nonlinear map* which is uniquely described by a *truth table* of three input variables (u_{i-1}, u_i, u_{i+1}). The choice of $\{-1, 1\}$ and not $\{0, 1\}$ as binary signals is crucial, because the state x_i and output y_i evolves in *real time* via a carefully designed *scalar ordinary differential equation*. According to this differential equation, the output y_i which is defined via an output equation $y_i = y(x_i)$ tends to either 1 or -1 after the solution x_i (with zero initial state $x_i(0) = 0$), reaches a *steady state*. In this way, the *attractors* of the *dynamical system* can be used to encode a *binary truth table*.

Aside from the *cell's intrinsic time scale* (which is of no concern in CA), an external clocking mechanism is introduced to reset the input u_i of each cell i at the end of each clock cycle by feeding back the steady state *output* $y_i \in \{-1, 1\}$, as an updated *input* $u_i \in \{-1, 1\}$, for the *next iteration*. This mechanism corresponds to the *periodic boundary condition* of the 1D cellular automaton in Fig. 1a.

Although CA are concerned only with the ring's evolutions over *discrete times*, any computer used to simulate CA is always a *continuous time system* with very small but non-zero time scale. Computers use transistors as devices, and each CA iteration involves the physical evolution of millions of transistors with its own $u_i \in \{-1, 1\}$, intrinsic dynamics. These transistors evolve in accordance with a large system of nonlinear differential equations governing the entire internal computer circuit and return the desired output after converging to their attractors in a non-zero amount of time.

These considerations lead us to the important result that, even in *discrete systems* like CA, there are *two different time scales* involved. The first one applies to the rule N while the second applies to the global patterns of evolution. To understand the complex dynamics of global patterns, it is necessary to analyze both time scales. By unfolding the truth tables of CA into an appropriate nonlinear dynamical system, we can exploit the theory of *nonlinear differential equations* to arrive at phenomena based on a precise mathematical theory, and not only on empirical observations.

For this purpose, we substituted the binary symbol 0 by the *real number* -1, and the input and output values 0 and 1 in the truth table of Fig. 1d by the real numbers -1 and 1, respectively. An advantage of working with the numeric rather than the symbolic truth table is the remarkable insights provided by the equivalent Boolean cube representation. Here, the eight vertices of the cube $(-1,-1,-1)$, $(-1,-1,1)$, $(-1,1,-1)$, $(-1,1,1)$, $(1,-1,-1)$, $(1,-1,1)$, $(1,1,-1)$ and $(1,1,1)$ are located exactly at the coordinates (u_{i-1}, u_i, u_{i+1}) of a *coordinate system* with the origin located at the center of the cube. The vertex $n = 0, 1, 2, \ldots, 7$ corresponding to row n of the truth table is coded blue if the output is -1, and red if the output is 1.

The choice of $\{-1, 1\}$ instead of $\{0, 1\}$ as binary signals is necessary, when the truth table is mapped onto a dynamical system where the states evolve in real time via an ordinary differential equation which is always based on the real number system. Each cell i is coupled only to its left neighbor cell $i - 1$ and right neighbor

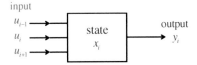

Fig. 3 Cell as dynamical system with state variable x_i, an output variable y_i, and three constant binary inputs u_{i-1}, u_i, and u_{i+1}

cell $i + 1$. As a dynamical system, each cell i has a *state variable* x_i, an *output variable* y_i, and three constant *binary inputs* u_{i-1}, u_i, and u_{i+1} (Fig. 3). Thus, the *dynamical system* is determined by a

$$\text{state equation}: \dot{x}_i = f(x_i; u_{i-1}, u_i, u_{i+1})$$
$$x(0) = 0 \,(\text{initital condition})$$
$$\text{output equation}: y_i = y(x_i).$$

Every CA can be mapped into a nonlinear dynamical system whose attractors encode precisely the associated truth table $N = 0, 1, 2, 3, \ldots, 255$. Function f models the *time-dependent change of states* and is defined by a scalar, ordinary differential equation of the form

$$\dot{x} = g(x_i) + w(u_{i-1}, u_i, u_{i+1}) \text{ with } g(x_i) \triangleq -x_i + |x_i + 1| - |x_i - 1|.$$

There are many possible choices of *nonlinear basis functions* for $g(x_i)$ and $w(u_{i-1}, u_i, u_{i+1})$. We have chosen the absolute value function $|x| = x$ for positive numbers x and $|x| = -x$ for negative numbers x as a nonlinear basis function, because the resulting equation can be expressed in an optimally compact form, and it allows us to derive the solution of the state equation in an explicit form. The scalar function $w(u_{i-1}, u_i, u_{i+1})$ can be chosen to be a composite function $w(\sigma)$ of a single variable $\sigma \triangleq b_1 u_{i-1} + b_2 u_i + b_3 u_{i+1}$ with $w(\sigma) \triangleq \{z_2 \pm |[z_1 \pm |z_0 + \sigma|]|\}$. This function is used to define the appropriate differential equation for generating the truth table of all 256 Boolean cubes. Thus, each rule of a cellular automaton corresponds to a particular set of *six real numbers* $\{z_0, z_1, z_2; b_1, b_2, b_3\}$, and *two integers* ± 1. Only 8 bits are needed to uniquely specify the differential equation associated with each rule N of a cellular automaton.

It can be proven that once the parameters defining a particular rule N are specified, then for any one of the eight inputs u_{i-1}, u_i, and u_{i+1} listed in the corresponding truth table of N, the solution x_i of the scalar differential equation will either increase monotonically from the initial state $x_i = 0$ towards a *positive equilibrium value* $\overline{x}_i(n) \geq 1$, henceforth denoted by attractor $Q_+(n)$, or decrease monotonically towards a *negative equilibrium state* $\overline{x}_i(n) \leq -1$, henceforth denoted by attractor $Q_-(n)$, when the input (u_{i-1}, u_i, u_{i+1}) is chosen from the coordinates of vertex n of the associated Boolean cube, or equivalently, from row n of the corresponding truth table, for $n = 0, 1, 2, \ldots, 7$ (Chua et al. 2002). Vertex n is painted red whenever its equilibrium value $\overline{x}_i(n) \geq 1$, and blue whenever $\overline{x}_i(n) \leq -1$. The color of all eight vertices for the associated *Boolean cube* will then be uniquely specified by the *equilibrium solutions* of the eight associated *differential equations*.

In general, we can summarize: once the parameters associated with a particular rule of a cellular automaton are specified, the corresponding *truth table* or *Boolean cube*, will be uniquely generated by the *scalar differential equation* alone. If the output equation of the dynamical system is $y_i = y(x_i) \triangleq \frac{1}{2}(|x_i + 1| - |x_i - 1|)$, then $y_i = +1$ when $x_i \geq 1$, and $y_i = -1$ when $x_i \leq -1$. The steady-state output at equilibrium is given explicitly by the formula $y_i = \text{sgn}\{\{w(\sigma)\}$ for any function $w(\sigma) \triangleq w(u_{i-1}, u_i, u_{i+1})$ with signum function $\text{sgn}(x) = +1$ for positive numbers x, $\text{sgn}(x) = -1$ for negative numbers x and $\text{sgn}(0) = 0$. For the particular $w(\sigma)$ in Fig. 4 the output (color) at equilibrium is given explicitly by the

$$\text{attractor color code}: y_i = \text{sgn}\{z_2 \pm |[z_1 \pm |z_0 + \sigma|]|\}.$$

Figure 4 contains four examples of dynamical systems and the rules they encode, each one identified by its rule number $N = 0, 1, 2, ..., 255$. The truth table for each rule N is generated by the associated dynamical system defined in the upper portion of each quadrant, and not from the truth table, thereby proving that each dynamical system and the rule of the cellular automaton that it encodes are one and the same. The truth table for each rule in Fig. 4 is cast in a format with only $2^{2^3} = 256$ distinct 1×3 neighborhood patterns. Each color picture consists of 30×61 pixels, generated by a 1D cellular automaton with 61 cells and a boundary condition with a specific rule N.

As an example, let us examine one of the rules from Fig. 4, rule 110, which will later on be identified as the simplest universal Turing machine known to date. With its differential equation, one can identify $\sigma = b_1 u_{i-1} + b_2 u_i + b_3 u_{i+1}$ with $b_1 = 1$, $b_2 = 2$, and $b_3 = -3$, and $w(\sigma) \triangleq \{z_2 \pm |[z_1 \pm |z_0 + \sigma|]|\}$ with $z_2 = -2$, $z_1 = 0$, and $z_0 = -1$. Thus, the attractor color code is explicitly given by $y_i = \text{sgn}[-2 + |u_{i-1} + 2u_i - 3u_{i+1} - 1)|]$.

Digital Dynamics with Difference Equations

The dynamics of dynamical systems are modeled with continuous differential equations. For computing the dynamics for digital CA, a program must use a "do loop" instruction which feeds back the output y_i^t of each cell at iteration t back to its inputs, to obtain the output y_i^{t+1} at the next iteration $t + 1$. Using the superscripts t and $t + 1$ as iteration number from one to the next generation, we can express each rule N explicitly in the form of a *nonlinear difference equation* with

$$u_i^{t+1} = \text{sgn}\{z_2 + c_2|[z_1 + c_1|(z_0 + b_1 u_{i-1}^t + b_2 u_i^t + b_3 u_{i+1}^t)|]|\},$$

where the *eight parameters* $\{z_0, z_1, z_2; b_1, b_2, b_3; c_1, c_2\}$ are given for each rule. Thus, the first main result of this chapter is that each of 256 1D CA that were studied by Steven Wolfram experimentally can be generated from a single scalar

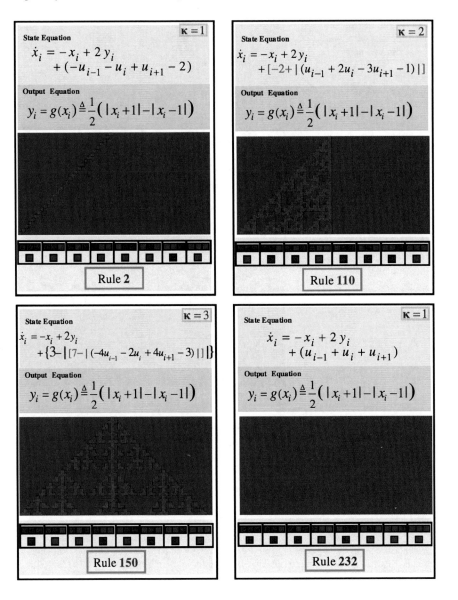

Fig. 4 Cellular automata with rules 2, 110, 150, 232 as dynamical systems. The initial condition is $x(0) = 0$

nonlinear differential equation or a corresponding nonlinear difference equation with at most eight parameters. These equation are also universal in the sense of a universal Turing machine (UTM), because we will later on see that at least one of the 256 rules (for example, rule 110) is capable of universal computation (Chua et al. 2003). For rule 110 (Fig. 5), we get $u_i^{t+1} = \text{sgn}\left(-2 + |u_{i-1}^t + 2u_i^t - \right.$

Fig. 5 Cellular automaton
as dynamical system with
difference equation

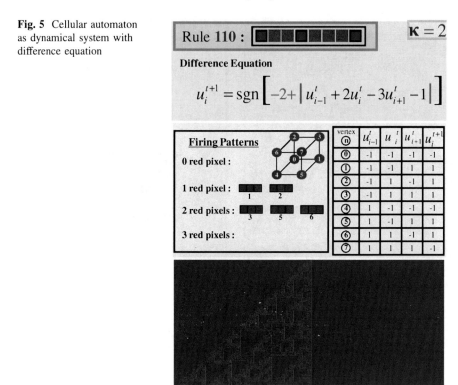

$3u_{i+1}^{t} - 1|)$. This kind of difference equation can be understood with elementary knowledge of basic mathematics, although it demonstrates important features of nonlinear dynamics.

References

L.O. Chua, S. Yoon, R. Dogaru, A nonlinear dynamics perspective of Wolfram's new kind of science Part I: Threshold of complexity. Int. J. Bifurcation Chaos (IJBC) **12**(12), 2655–2766 (2002)

L.O. Chua, V.I. Sbitnev, S. Yoon, A nonlinear dynamics perspective of Wolfram's new kind of science Part II: Universal neuron. Int. J. Bifurcation Chaos (IJBC) **13**(9), 2377–2491 (2003)

S. Wolfram, *A New Kind of Science* (Wolfram Media Inc, Champaign, 2002)

Chapter 3
Complexity in the Universe of Cellular Automata

The colored toy cubes contain all the information about the complex dynamics of cellular automata. An important advantage of the Boolean cube representation is that it allows us to define an index of complexity (Chua et al. 2002). Each one of the 256 cubes is obviously characterized by different clusters of red or blues vertices which can be separated by parallel planes. On the other hand, the separating planes can be analytically defined in the coordinate system of the Boolean cubes. Therefore, the complexity index of a cellular automaton with local rule N is defined by the minimum number of parallel planes needed to separate the red vertices of the corresponding Boolean cube N from the blue vertices. Figure 1 shows three examples of Boolean cubes for the three possible complexity indices $\kappa = 1, 2, 3$ with one, two and three separating parallel planes. There are 104 local rules with complexity index $\kappa = 1$. Similarly, there are 126 local rules with complexity index $\kappa = 2$, and only 26 local rules with complexity index $\kappa = 3$. This analytically defined complexity index is to be distinguished from Wolfram's complexity index based on phenomenological estimations of pattern formation.

Complexity Index of Cellular Automata

In the context of colored cubes of cellular automata, *separability* refers to the number of cutting (parallel) planes separating the vertices into clusters of the same color. For rule 110, for example, we can introduce two separating parallel planes in the corresponding colored cube, and which are distinguished in Fig. 1b by two different colors: The red vertices 2 and 6 lie above a yellow plane. The blue vertices 0, 4, and 7 lie between the yellow and a light blue plane. The red vertices 3, 1, and 5 lie below the light blue plane. It is well-known that the cellular automaton of rule 110 is one of the few types of the 256 automata which are *universal Turing machines*. In the sense of Wolfram's class 3 of computer experiments, it produces very complex patterns (Wolfram 2002).

K. Mainzer and L. Chua, *The Universe as Automaton*, SpringerBriefs
in Complexity, DOI: 10.1007/978-3-642-23477-4_3, © The Author(s) 2012

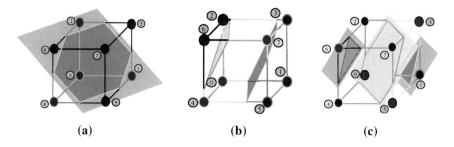

Fig. 1 Examples of complexity index $\kappa = 1$, 2, 3 with parallel planes that separate all vertices having one color on one side from those having a different color on the other side, for rule 232 (**a**), rule 110 (**b**), and rule 150 (**c**)

An example of an automaton which can only produce very simple patterns is rule 232. There is only *one separating plane* cutting the corresponding Boolean cube for separating colored points (Fig. 1a): Red vertices 3, 5, 6, and 7 lie above a light blue plane. The blue vertices 0, 1, 2, and 4 lie below the light blue plane. A colored Boolean cube with *three parallel separating planes* is shown in Fig. 1c, representing the cellular automaton of rule 150: The blue vertex 6 lies above a green plane. The red vertices 2, 4, and 7 lie between a yellow plane and the green plane. The blue vertices 0, 3, and 5 lie between the yellow plane and a light blue plane. The blue vertex 1 lies below the light blue plane. Obviously, it is not possible to separate the 8 vertices into three colored clusters and at the same time separate them by two parallel planes, no matter how the planes are positioned.

A rule whose colored vertices can be separated by only one plane is said to be *linearly separable*. An examination of the 256 Boolean cubes shows that 104 among them are linearly separable. The remaining 152 rules are not linearly separable. Each rule can be separated by various numbers of parallel planes. In general, there is a unique integer κ, henceforth called the *complexity index* of rule N, which characterizes the geometrical structure of the corresponding Boolean cube, namely the minimum number of parallel planes that are necessary to separate the colored vertices. All linearly separable rules have a *complexity index* $\kappa = 1$. An analysis of the remaining 152 linearly non-separable rules shows that they have a complexity index of either 2 or 3. For example, rule 110 has a complexity index $\kappa = 2$, whereas rule 150 has a complexity index $\kappa = 3$. No rule with complexity index $\kappa = 1$ is capable for generating complex patterns, even for random initial conditions. The emergence of complex phenomena significantly depends on a minimum complexity of $\kappa = 2$. In this sense, complexity index 2 can be considered the threshold of complexity for 1-dimensional cellular automata.

Analytical Geometry of Boolean Cubes

Our complexity index of cellular automata followed from the symmetries of their Boolean cubes. The geometrical introduction of separating planes can easily be understood as a complexity index for colored cubes. Contrary to Steven Wolfram's quasi-empirical index of patterns that are intuitively more-or-less complex, our complexity index can be mathematically defined in *analytical geometry*. The separating planes of a Boolean cube are determined by the nonlinear function $w(u_{i-l}, u_i, u_{i+l})$ of the corresponding state equation. Geometrically, it is interpreted as a *scalar function* $w(\sigma)$ of only one variable $\sigma \triangleq b_1 u_{i-1} + b_2 u_i + b_3 u_{i+1}$, representing an axis in the *coordinate system* (u_{i-1}, u_i, u_{i+1}) with orientation b_1, b_2, and b_3 (Chua et al. 2002). Each colored vertex of a Boolean cube can be mapped on the σ-axis by a perpendicular projection. If we plot the curve of $w(\sigma)$ on the σ-axis, we observe that its zero-crossing points, σ_o with $w(\sigma_o) = 0$, separate the colored points on the projection-axis into clusters of common color. Thus, $w(u_{i-1}, u_i, u_{i+1})$ is sometimes called the *discriminant function*. Each zero-crossing point of $w(\sigma)$ defines a 2-*dimensional plane* $\sigma_0 = b_l u_{i-l} + b_2 u_i + b_3 u_{i+l}$ in the 3-*dimensional coordinate system* (u_{i-1}, u_i, u_{i+1}), separating the colored vertices of a Boolean cube into clusters of common color (Fig. 1).

If the colored vertices can be separated by only one plane, we called the corresponding Boolean rule linearly separable. The reason is now obvious, because in this case the associated discriminant function $w(\sigma)$ is a straight line. In the cases of several separating planes, there are several zero-crossing points of the separating curve associated with a nonlinear discriminant function $w(\sigma)$. Then, the corresponding Boolean rule was called linearly non-separable. The projection technique delivers a precise procedure for computing the complexity indices of Boolean rules and their associated 1-dimensional cellular automata.

From Simple Building Blocks to Complex Compositions

Linearly separable local rules have a complexity index $\kappa = 1$. In general, they are the *simplest building blocks* of Boolean functions of any dimension. From an engineering point of view, they are also the simplest ones to implement to a chip. A technical advantage is that linearly separable rules are the fastest to execute on a chip. They only need a few nanoseconds with silicon technology, and operate at the speed of light with optical technology. The speed of the associated cellular automata is independent of the size of the array.

All 152 linearly non-separable rules with three inputs can be *decomposed* in terms of at most three linearly separable rules, and combined pixelwise via only AND and OR logic operations. Figure 2 shows examples of decompositions for rule 110 involving only one AND operation (a) and for rule 105 with one AND and one OR operation (b). Rule 105 is thus one of the most complicated 1-dimensional

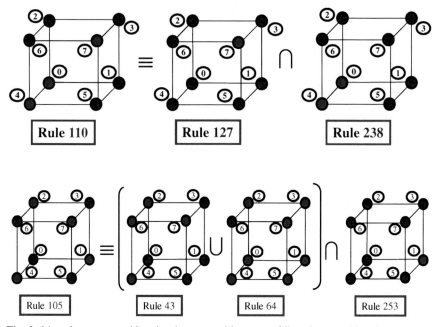

Fig. 2 Linearly non-separable rules decomposed in terms of linearly separable rules

cellular automata to implement on a chip. Concerning dynamic complexity, rule 105 has also the highest complexity index 3.

All 256 Boolean cubes can be classified into *equivalence classes* with *identical complexity index* (Chua et al. 2004). The corresponding Boolean rules are called equivalent iff there exists a transformation, mapping the one rule onto the other one, and vice versa. In the case of a Red↔Blue complementary transformation, the colors of the corresponding vertices of corresponding Boolean cubes (Fig. 1) are the complement of each other, i.e., corresponding red vertices become blue, and vice versa. In the case of a Left–Right symmetrical transformation, the colors between vertex 3 and 6, as well as between vertex 1 and 4 in one Boolean cube (Fig. 1) are interchanged in order to get the other one. Obviously, rule 150 (Fig. 1c) is invariant under a Left–Right symmetrical transformation, because vertex 1 and 4 have identical colors (red), as well as vertex 3 and 6 (blue). All members belonging to the same *global* equivalence classes of Boolean rules have an identical complexity index and show dynamic behavior, which can be predicted from one another. Thus, it is sufficient to study only one representative member of each *global* equivalence class. In general, 38 independent *linearly-separable* rules with complexity index $\kappa = 1$ and 50 independent *linearly non-separable* rules (41 examples with complexity index $\kappa = 2$, and 9 examples with complexity index $\kappa = 3$) can be identified. It follows that the nonlinear dynamics and dynamic complexity of the 256 Boolean functions with three binary inputs are completely characterized by only 88 independent representatives.

Obviously, index 2 is a *threshold of complexity*. No local rule with complexity index $\kappa = 1$ is capable of generating complex patterns. It is clear, therefore, that to exhibit emergence and complex phenomena, a local rule must have a minimum complexity index of $\kappa = 2$. This analytically-based concept of complexity is certainly consistent with the following empirically-based observations of Steven Wolfram (Wolfram 2002): *The examples in this and the previous chapter suggest that if the rules for a particular system are sufficiently simple, then the system will only ever exhibit purely repetitive behavior. If the rules are slightly more complicated, then nesting will also often appear. But to get complexity in the overall behavior of a system, one needs to go beyond some threshold in the complexity of its underlying rules. In any case, it ultimately takes only simple local rules to produce global patterns of great complexity.*

Computational Complexity and Universal Computability

A motivation for the introduction of a complexity index is also *computational complexity*. The class of cellular automata with complexity index $\kappa = 2$ contains examples with universal computation (e.g., $N = 110$), but the local rules with complexity index $\kappa = 1$ are not capable of universal computation. It follows that $\kappa = 2$ also represents a threshold of computational complexity.

Universal computation is a remarkable concept of computational complexity which dates back to Alan Turing's universal machine (Turing 1936/1937). Universal cellular automata are well-known since Conway's Game of Life (Martin 1994; Rendell 2002). A *universal Turing machine* can, by definition, simulate any Turing machine. According to the Church-Turing thesis, any algorithm or effective procedure can be realized by a Turing machine. Now Turing's famous *Halting problem* comes in. Following his proof, there is no algorithm that can decide for an arbitrary computer program and initial condition, whether or not it will stop in the long run. (A computer program cannot stop if it must follow a closed loop.) Consequently, for a system with universal computation (in the sense of a universal Turing machine), we cannot predict if it will stop in the long run or not. Assume that we were able to do that. Then, in the case of a universal Turing machine, we could also decide whether any Turing machine (which can be simulated by the universal machine) would stop or not. That is obviously a contradiction to Turing's result of the *Halting problem*. Thus, systems with universal computation are unpredictable.

Unpredictability is obviously associated with a high degree of complexity. It is extremely surprising that systems with simple rules of behavior like cellular automata lead to complex dynamics which are no longer predictable. We will be very curious to discover examples of these, in principle, unpredictable automata, in nature.

References

L.O. Chua, S. Yoon, R. Dogaru, A nonlinear dynamics perspective of Wolfram's new kind of science. Part I: Threshold of complexity. Int. J. Bifurcation Chaos (IJBC) **12**(12), 2655–2766 (2002)

L.O. Chua, V.I. Sbitnev, S. Yoon, A nonlinear dynamics perspective of Wolfram's new kind of science. Part III: Predicting the unpredictable. Int. J. Bifurcation Chaos (IJBC) **14**, 3689–3820 (2004)

B. Martin, A universal cellular automaton in quasi-linear time and its S-m-n form. Theor. Comput. Sci. **123**, 199–237 (1994)

P. Rendell, A Turing machine in Conway's game of life, extendable to a universal Turing machine, in *Collision-Based Computing*, ed. by A. Adamatzky (Springer, New York, 2002)

A.M. Turing, On computable numbers with an application to the Entscheidungsproblem. Proc. Lond. Math. Soc. **42**(2), 230–265, corrections, ibid, **43**, 544–546 (1936/1937)

S. Wolfram, *A New Kind of Science* (Wolfram Media, Champaign, 2002)

Chapter 4
Symmetry in the Universe of Cellular Automata

A cursory inspection of the discrete time evolutions of the 256 local rules reveals some similarity and partial symmetry among various evolved patterns. It reminds us of more-or-less random observations in the natural sciences that demand the unification of mathematical explanations with fundamental laws. The unifying theory of physics is based on the assumption of fundamental mathematical symmetries (Mainzer 1996, 2005). According to this view, the variety and complexity of natural phenomena have evolved from just a few principles of symmetry. They are the *"Holy Grail"* of the Universe which is sought by prominent scientists and research groups all over the world. For the universe of cellular automata, we found the fundamental symmetries in the gallery of Boolean cubes (Chua et al. 2004). Thus, at least in the toy world of cellular automata, the importance of symmetry laws can easily be imagined and understood.

Local Equivalence of Cellular Automata

But even in the universe of cellular automata, the situation is sophisticated. The Boolean cubes of many different pairs of local rules seem to be related by some *symmetry transformations*, such as complementation of the vertex colors (examples being rules 145 and 110). Yet, their evolved patterns are so different that it is impossible to relate them. How do we make sense of all these observations? In the case of rule 145 and 110, the associated Boolean cubes are related by a "red ↔ blue vertex transformation". This is denoted as a *local complementation operation* T^C, because complementation is locally restricted. Intuitively, one might expect that their respective evolved patterns must also be related by a global complementation operation. But upon comparing the two evolved patterns (Fig. 1), the intuition turns out to be generally wrong. It is only true in a local sense with respect to special iterations. For example, starting from the same initial pattern

K. Mainzer and L. Chua, *The Universe as Automaton*, SpringerBriefs
in Complexity, DOI: 10.1007/978-3-642-23477-4_4, © The Author(s) 2012

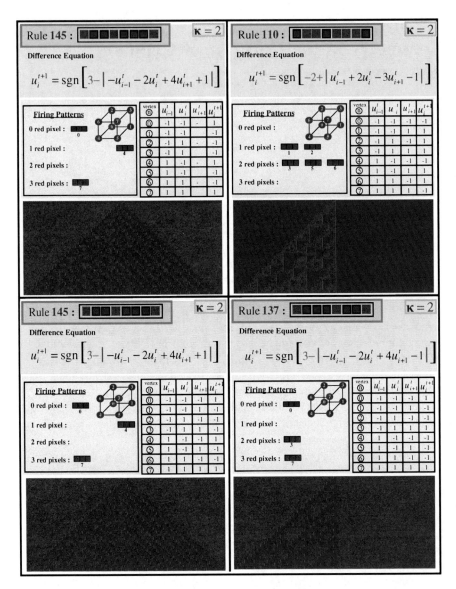

Fig. 1 The evolution of rules 110 and 145 only reveal a local complement relationship in the first iteration, but 110 and 137 reveal global symmetrical relationship

(single red center pixel) in the first row, we find the output (first iteration) of rule 145 is in fact the complement of that of rule 110; namely, two blue pixels for 145 and two red pixels for 110 at corresponding locations to the left of center. All other pixels at corresponding locations are also complements of each other.

However, the next iteration (row 3) under rules 145 and 110 in Fig. 1 are not complements of each other. The reason is that unlike the initial input u_i^0, $i = 0, 1, 2,$ ..., n, which is the same for both 145 and 110, the next input u_i^1, $i = 0, 1, 2, ..., n$ (for $t = 1$ in row 2), needed to find the next iteration (row 3) is different and there is no reason for the outputs u_i^2 (for $t = 2$) at corresponding locations to be the complement of each other. In these cases, a pair of local rules is equivalent only in a local sense with respect to *"local in iteration time"*, and not local in the usual sense of a spatial neighborhood.

In general, we define

Local Equivalence: Two local rules N and N' are said to be *locally equivalent* under a transformation $\mathbf{T}$: $N \rightarrow N'$ iff the output u_i^1 of N after one iteration of any initial input pattern u_i^0 can be found by applying the transformed input $\mathbf{T}(u_i^0)$ to rule N' and then followed by applying the *inverse transformation* $\mathbf{T}^{-1}$: $N' \rightarrow N$ to u_i^1.

Global Equivalence of Cellular Automata

Global aspects can be observed in the evolved patterns for the rules 110, 137 (Fig. 1), 124, and 193 (Fig. 2). Despite the fact that the respective Boolean cubes of these three rules do not seem to be related in an obvious way, their output patterns are so precisely related that one could predict the evolved pattern over all times t for each local rule 110, 124, 137, and 193. For example, the evolved output pattern of rule 124 can be obtained by a reflection of that of 110 about the center line, namely a bilateral transformation. The output of rule 193 can be obtained by applying the complement of u_i^0 (i.e. blue center pixels amidst a red background) to rule 110 and then taking the complement of the evolved pattern from 110. The output of rule 137 can be obtained by repeating the above algorithm for 193, and then followed further by a reflection about the center line. It can be proved that these algorithms remain valid for all initial input patterns. This result is most remarkable because it allows us to *predict* the evolved patterns from arbitrary initial configurations of three rules over all iterations, and not just for one iteration as in the case of local equivalence.

In general, we define

Global Equivalence: Two local rules N and N' are said to be *globally equivalent* under a transformation $\mathbf{T}$: $N \rightarrow N'$ iff the output u_i^t of N can be found, for any t, by applying the transformed input $\mathbf{T}(u_i^0)$ to local rule N' and then followed by applying the *inverse transformation* $\mathbf{T}^{-1}$: $N' \rightarrow N$ to u_i^t, for any $t = 1, 2, ...$.

Obviously, the four rules 110, 124, 137, and 193 are globally equivalent in the sense that the evolved patterns of any three members of this class can be trivially predicted from the fourth, for all iterations. Therefore, these four rules have identical nonlinear dynamics for all initial input patterns and therefore they represent only one generic rule, henceforth called a *global equivalence class*. This global property is not only true for four rules, but also for all rules, thereby allowing us to partition the 256 rules into only 88 global equivalence classes. It is convenient to identify these equivalence classes with the symbol ε_m^κ, where κ is the

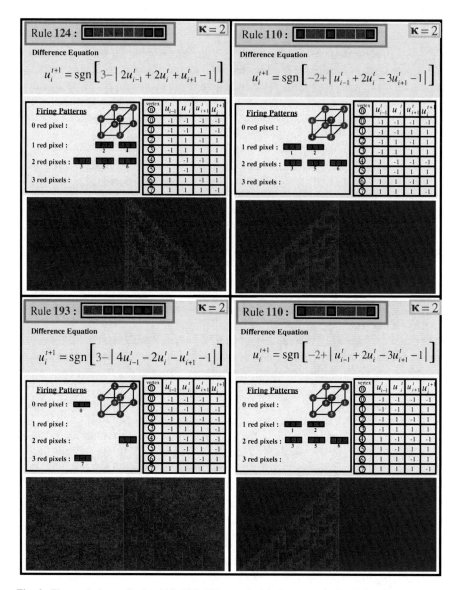

Fig. 2 The evolutions of rules 110, 124, 193 reveal global symmetrical relationships

complexity index and *m* the *class number*. There are 38 cellular automata belonging to the equivalence classes ε_m^1 with complexity index $\kappa = 1$ and $m = 1$, 2, ..., 38. The equivalence classes ε_m^2 with complexity index $\kappa = 2$ are distinguished by $m = 1, 2, ..., 41$. In addition, there are nine global equivalence classes with complexity index $\kappa = 3$. They are identified by ε_m^3 with $m = 1, 2, ..., 9$.

This result is significant because it asserts that one only needs to study in depth the dynamics and long-term behavior of 88 representative local rules. Moreover, since 38 of these 88 dynamically distinct rules have complexity index $\kappa = 1$, and are therefore trivial, we are left with only 50 local rules (41 rules with $\kappa = 2$ and 9 rules with $\kappa = 3$) that justify further in-depth investigation.

Symmetry with Global Transformations

It can be proven that every local rule belongs to a global equivalence class determined by certain global transformations. There are three global transformations, namely, *global complementation* $\overline{T}$, *left–right complementation* $T^\star$ and *left–right transformation* $T^\dagger$ which are distinguished as *symmetry transformations* in the universe of cellular automata. The four rules 110, 124, 137, and 193 are globally equivalent to each other in the sense that their long term dynamics (as $t \rightarrow \infty$) are mathematically identical with respect to the three global transformations $T^\dagger$, $T^\star$ and $\overline{T}$.

The intuitive meaning of these symmetry transformations can easily be seen in Fig. 3. In this picture, all four patterns of rules 110, 124, 137, and 193 have 60 rows corresponding to iterations numbers $t = 0, 1, 2, ..., 59$, and 61 columns, corresponding to 61 cells ($n = 60$). All patterns have a random initial condition ($t = 0$), or its reflection, complementation, or both. The two patterns 124 and 110 at the top are generated by a left–right transformation $T^\dagger$, and are related by a bilateral reflection about an imaginary vertical line situated midway between the two patterns. The two patterns 193 and 137 below are likewise related via $T^\dagger$, and exhibit the same bilateral reflection symmetry. The two vertically situated local rules 137 and 110, as well as 193 and 124 are related by a global complementation $\overline{T}$. The two diagonally-situated local rules 124 and 137, as well as 193 and 110 are related by a left–right complementation $T^\star$.

The *geometrical definition* of these symmetry transformations is easy to understand and can even be imagined with the help of our toy cubes of cellular automata. Mathematically, these transformations are defined by 3×3 *matrices* $\overline{T}_u$, $T_u^\star$ and $T_u^\dagger$. Each of the three matrices transforms the three axes (u_{i-1}, u_i, u_{i+1}), drawn through the center of the *Boolean cube* into a transformed set of axes (u'_{i-1}, u'_i, u'_{i+1}). These matrix representations also only need basic mathematics.

Symmetry of Left–Right Transformation $T^\dagger$

We can easily imagine how $T^\dagger$ implements a mirror reflection of a Boolean cube about a diagonal plane formed by the 4 vertices 0, 2, 5, 7 of the Boolean cube (Fig. 4). Geometrically, the *left–right transformation matrix* $T_u^\dagger$ switches

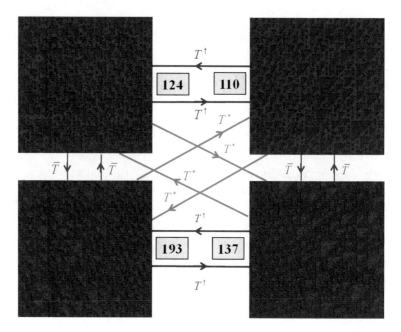

Fig. 3 Global equivalence of rules 110, 124, 137, and 193

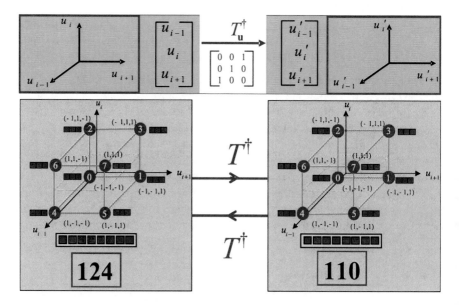

Fig. 4 Left–right transformation

the two horizontal axes u_{i-1} and u_{i+1}. This means that operating on a *Boolean cube N* by $\mathbf{T}_u{}^\dagger$ is equivalent to switching the two pairs of vertices 4 and 6 on the left and 1 and 3 on the right in the Boolean cube N to obtain a transformed Boolean cube $N' = \mathbf{T}_u{}^\dagger(N)$. An example of this simple transformation is $110 = \mathbf{T}_u{}^\dagger(124)$ for rule $N = 124$. In matrix notation, the transformation is realized by

$$\begin{bmatrix} 0 & 0 & 1 \\ 0 & 1 & 0 \\ 1 & 0 & 0 \end{bmatrix} \begin{bmatrix} u_{i-1} \\ u \\ u_{i+1} \end{bmatrix} = \begin{bmatrix} u_{i+1} \\ u_i \\ u_{i-1} \end{bmatrix} \text{ with the transformed values } \begin{aligned} u'_{i-1} &= u_{i+1} \\ u'_i &= u_i \\ u'_{i+1} &= u_{i-1} \end{aligned} \ .$$

Since $\left(\mathbf{T}_u^\dagger\right)^{-1} = \mathbf{T}_u^\dagger$, the transformation is its own *inverse*. Consequently, 124 $= \mathbf{T}_u^\dagger(110)$. An examination of the corresponding Boolean cubes reveals that the left–right transformation $\mathbf{T}_u^\dagger$ is equivalent to a reflection of the Boolean cube about the diagonal plane passing through vertices 0, 5, 7, and 2. This *bilateral (mirror) symmetry* can easily be imagined. The corresponding *global left–right transformation* $\mathbf{T}^\dagger$ (without the subscript u) is defined by augmenting $\mathbf{T}_u^\dagger$ with the output variable y_i, which represents the color of the vertices as follows:

$$\begin{bmatrix} 0 & 0 & 1 & 0 \\ 0 & 1 & 0 & 0 \\ 1 & 0 & 0 & 0 \\ 0 & 0 & 0 & 1 \end{bmatrix} \begin{bmatrix} u_{i-1} \\ u_i \\ u_{i+1} \\ y_i \end{bmatrix} = \begin{bmatrix} u_{i+1} \\ u_i \\ u_{i-1} \\ y_i \end{bmatrix} \text{ with the transformed values } \begin{aligned} u'_{i-1} &= u_{i+1} \\ u'_i &= u_i \\ u'_{i+1} &= u_{i-1} \\ y_i &= y_i \end{aligned} \ .$$

Symmetry of Global Complementation $\overline{\mathbf{T}}$

The reader can also easily imagine how $\overline{\mathbf{T}}$ implements a point reflection of a Boolean cube about the center of the cube, followed by complementing the color of each vertex (Fig. 5). Geometrically, *global complementation* $\overline{\mathbf{T}}_u$ means that transforming a *Boolean cube N* by $\overline{\mathbf{T}}_u$ is equivalent to switching the four pairs of vertices $\{0, 7\}$, $\{1, 6\}$, $\{2, 5\}$, and $\{3, 4\}$ located along the four imaginary diagonal lines through the center of the Boolean cube. Why is the output pattern of rule 145 that was mentioned earlier (Fig. 1) not the complement of that of 110 when the colors of the corresponding vertices of 145 and 110 are complements of each other? Obviously, we must not only switch the diagonal vertex pairs, as above, but must also follow this switch with the intermediate operation $\overline{\mathbf{T}}_u$ by changing the colors of the vertices to their complementary colors. It follows that the *global complementation* $\overline{\mathbf{T}}$ must be defined as follows:

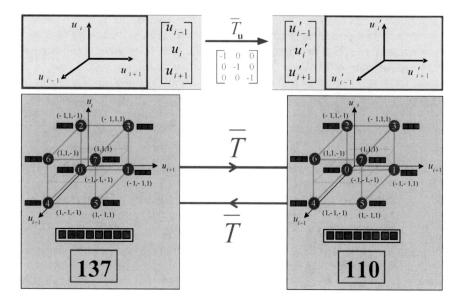

Fig. 5 Global complementation

$$
\begin{bmatrix}
-1 & 0 & 0 & 0 \\
0 & -1 & 0 & 0 \\
0 & 0 & -1 & 0 \\
0 & 0 & 0 & -1
\end{bmatrix}
\begin{bmatrix}
u_{i-1} \\
u_i \\
u \\
y_i
\end{bmatrix}
=
\begin{bmatrix}
-u_{i-1} \\
-u_i \\
-u_{i+1} \\
-y_i
\end{bmatrix}
$$

with the transformed values
$$
\begin{aligned}
u'_{i-1} &= -u_{i-1} \\
u'_i &= -u_i \\
u'_{i+1} &= -u_{i+1} \\
y'_i &= -y_i
\end{aligned}
$$

The complement operation in row 4 of $\overline{\mathbf{T}}$ is equivalent to applying the *local red–blue complementation operator* $\mathbf{T}^C$ that was introduced before.

Symmetry of Left–Right Complementation $\mathbf{T}^\star$

In geometrical interpretation, the *left–right complementation* $\mathbf{T}_u^\star$ (Fig. 6) is equivalent to the composition of two operations $\mathbf{T}_u^\dagger$ and $\overline{\mathbf{T}}_u$. To obtain the *global left–right complementation* $\mathbf{T}^\star$ requires that we follow up the above composition $\mathbf{T}_u^\star = \overline{\mathbf{T}}_u \cdot \mathbf{T}_u^\dagger$ by a *local complementation* $\mathbf{T}^C$ as follows:

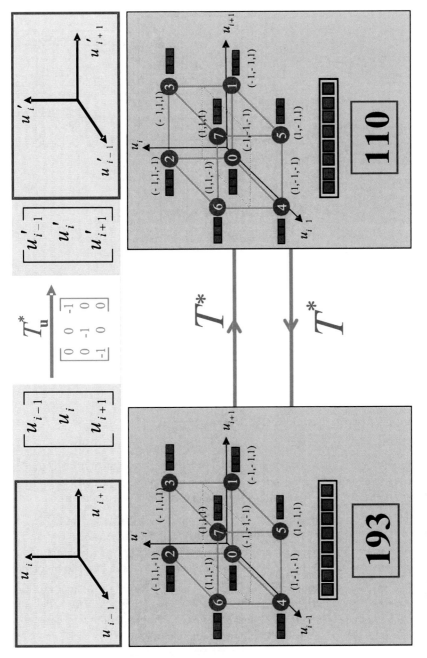

Fig. 6 Left–right complementation

Fig. 7 Global symmetry and Klein's Vierergruppe

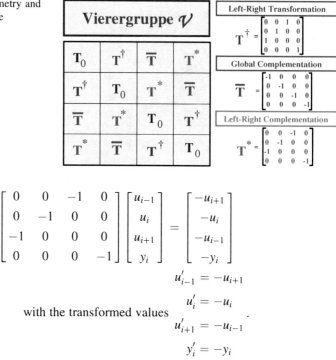

$$\begin{bmatrix} 0 & 0 & -1 & 0 \\ 0 & -1 & 0 & 0 \\ -1 & 0 & 0 & 0 \\ 0 & 0 & 0 & -1 \end{bmatrix} \begin{bmatrix} u_{i-1} \\ u_i \\ u_{i+1} \\ y_i \end{bmatrix} = \begin{bmatrix} -u_{i+1} \\ -u_i \\ -u_{i-1} \\ -y_i \end{bmatrix}$$

with the transformed values

$$u'_{i-1} = -u_{i+1}$$
$$u'_i = -u_i$$
$$u'_{i+1} = -u_{i-1}$$
$$y'_i = -y_i$$

For example, applying the local complementation $\mathbf{T}^C$ to 145 results in the desired rule 110 (*see* Fig. 1). Again, it is $(\mathbf{T^\star})^{-1} = \mathbf{T^\star}$.

Global Symmetry of Klein's Vierergruppe $\mathcal{V}$

The three global transformations $\mathbf{T}^\dagger$, $\mathbf{T^\star}$ and $\overline{\mathbf{T}}$ are generated from elements of the classic *noncyclic four-element Abelian group* $\mathcal{V}$, originally called the *"Vierergruppe"* by the German mathematician Felix Klein (Speiser 1956). The four elements of $\mathcal{V}$ are constructed from the 3×3 matrices $\mathbf{T_0}$, $\overline{\mathbf{T}}_u$, $\mathbf{T}_u{}^\star$ and $\mathbf{T}_u^\dagger$. The symbol $\mathbf{T_0}$ denotes the identity, or unit matrix, of any dimension. The actual transformations, however, that allow us to establish the long-term correlations among members of each of the 88 *global equivalence classes* of all 256 cellular automata are the 4×4 matrices $\mathbf{T_0}$, $\mathbf{T}^\dagger$, $\mathbf{T^\star}$ and $\overline{\mathbf{T}}$. Figure 7 shows that they are related by the group multiplication table of Klein's Vierergruppe $\mathcal{V}$. This is the only abstract mathematical group which makes it possible to predict the *long-term correlations* among all members of the four remarkable rules 110, 124, 137, and 193.

These results are global in the sense of asymptotic time behavior as $t \to \infty$. It proves that even though there are 256 distinct local rules of 1-dimensional

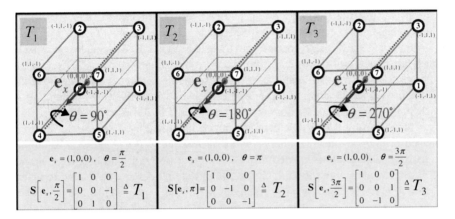

Fig. 8 Rotation matrices T_k for rotating a Boolean cube by $\theta°$ around some axis, such that the rotated cube is indistinguishable (except for the color of the vertices) from the original cube

cellular automata, there are only 88 distinct global behaviors, a fundamental result predicted by the identification of 88 global equivalence classes ε_m^K.

Local Symmetry Classes of Cellular Automata

How can *local equivalence classes* be distinguished in the universe of cellular automata and what are they good for? Local equivalence refers to the color of vertices in a Boolean cube of a cellular automaton. There are 23 distinct ways to rotate the *Boolean cube* about some axis by a certain angle so that the rotated cube coincides with the original cube, except for the color of their vertices.

Each *rotation operation* can be defined mathematically as a transformation of the axes (u_{i-1}, u_i, u_{i+1}) onto the rotated axes $(u'_{i-1}, u'_i, u'_{i+1})$ via a 3×3 matrix:

$$\begin{bmatrix} u'_{i-1} \\ u'_i \\ u'_{i+1} \end{bmatrix} = [T_k] \begin{bmatrix} u_{i-1} \\ u_i \\ u_{i+1} \end{bmatrix}, \text{ where } k = 1, 2, \ldots, 23.$$

The *transformation matrix* T_k is given for each rotation operation $S[e_k, \theta]$, where e_k denotes the rotation axis, and where x, y, z codes for u_{i-1}, u_i, and u_{i+1}. In Fig. 8, the first three transformations illustrate the situation. The rotation axis is shown by bold red arrows and labeled accordingly, including the rotation angle θ.

All 24 rotation matrices $T_0, T_1, \ldots, T_{23}$ including the identity matrix T_0, form a 24-element *non-Abelian* (non-commutative) *group* **R** with 13 *subgroups* (Hamermesh 1962). The first subgroup generates the three rotation matrices T_1 to T_3, the second subgroup generates the three rotation matrices T_4 to T_6, and the third subgroup generates the three rotation matrices T_7 to T_9. The next six subgroups generate the six rotation matrices T_{10} to T_{15}. The last four subgroups generate the eight remaining rotation matrices T_{16} to T_{23}. Every rotation matrix T_k, with $k = 0$, 1, 2, …, 23, can be generated by repeated compositions of only T_1 and T_4.

Although each of the 23 rotation matrices T_k maps a Boolean cube onto itself, the rotated cube will be different in general from the unrotated cube, because the color red or blue may not match at the corresponding vertices. However, there are 30 *distinct subsets* among the 256 Boolean cubes where the cubes belonging to each subset have matched color vertices after rotating each cube by an appropriate T_k. These 30 subsets of Boolean cubes can be identified by the symbol S_m^n, where the subscript m denotes the *number of red vertices of all Boolean cubes* belonging to map S_m^n. Just as a left-handed glove is different from a right-handed glove, not all Boolean cubes with the same number m of red vertices can be rotated to match each other. Each subset of m red vertices which do match would therefore constitute a separate class. The superscript n of S_m^n therefore denotes the subset number, henceforth called the *chiral number*, a generalization of the chemist's terminology, which was introduced for similar purposes in classifying molecules.

It is remarkable that all local rules belonging to the same subset S_m^n must have the same *complexity index* κ. Notice that the complexity index κ is the minimum number of parallel planes needed to separate the red vertices of Boolean cube N from the blue vertices. There is a correlation of κ with the minimum number α of *absolute-value functions* required by the Boolean output equation. In particular, the number $\alpha + 1$ is exactly equal to the minimum number of parallel separating planes. Hence, all local rules with the complexity index $\kappa = 1$ can be generated from the following equation with an absolute-value function of $\alpha = 0$:
$\kappa = 1$ local rules:

$$u_i^{t+1} = \mathrm{sgn}\{z_0 + b_1 u_{i-1}^t + b_2 u_i^t + b_3 u_{i+1}^t\}.$$

There are 104 local rules with complexity index $\kappa = 1$. They correspond to the classes of all *linearly-separable* Boolean functions. Similarly, there are 126 local rules with complexity index $\kappa = 2$. They can be generated from the following nonlinear equation with just an absolute-value function of $\alpha = 1$:
$\kappa = 2$ local rules:

$$u_i^{t+1} = \mathrm{sgn}\{z_1 + c_1 |(z_0 + b_1 u_{i-1}^t + b_2 u_i^t + b_3 u_{i+1}^t)|\}.$$

Only local rules with complexity index $\kappa = 3$ require two absolute-value functions, i.e., $\kappa = 2$. Therefore, it follows from the definition of the complexity index that all local rules belonging to the same subset S_m^n must have the same index κ.

The Holy Grail of Symmetry and Computability

Because the local rule 110 has been proved to be capable of *universal computation*, it follows that all four local rules of the Vierergruppe $\mathcal{V}$ are *universal Turing machines*. The fundamental importance of the universality result was being able to

Fig. 9 Universal symmetry and computability in the universe of cellular automata

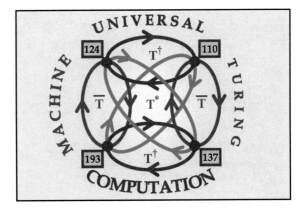

exploit the *symmetry* of the Boolean cubes to identify equivalence classes among the 256 rules. The discovery of the *Vierergruppe* $\mathcal{V}$ and the *rotation group* **R** has led to the major logical classifications of the 256 local rules into 88 *global equivalence classes* ε_m^κ and 30 *local equivalence classes* S_m^κ. The significance of the 88 global equivalence classes ε_m^κ is similar to the classification of computational algorithms into various complexity classes, for example, the *N*- or *NP*-classes, in the sense that any property that applies to one member of ε_m^κ applies to the other members in the same global equivalence class.

The *universality* of the four rules 110, 124, 137, and 193 and their identical long-term dynamic behaviors, with respect to the symmetry transformations of the Vierergruppe $\mathcal{V}$, are encapsulated in the commutative diagram shown in Fig. 9. Thus, Klein's Vierergruppe represents the fundamental symmetry law of the 256 two-state one-dimensional cellular automata. It is the *"Holy Grail"* of a unified theory in the universe of these cellular automata, containing all information about their nonlinear dynamics.

References

L.O. Chua, V.I. Sbitnev, S. Yoon, A nonlinear dynamics perspective of Wolfram's new kind of science: Part III: Predicting the unpredictable. Int. J. Bifurc. Chaos **14**, 3689–3820 (2004)

M. Hamermesh, *Group Theory and its Applications to Physical Problems* (Addison-Wesley, Reading, 1962)

K. Mainzer, *Symmetries of Nature* (De Gruyter, New York, 1996) (German 1988: Symmetrien der Natur. De Gruyter, Berlin)

K. Mainzer, *Symmetry and Complexity: The Spirit and Beauty of Nonlinear Science* (World Scientific, Singapore, 2005)

A. Speiser, *Die Theorie der Gruppen von endlicher Ordnung*, 4th edn. (Birkhäuser, Basel, 1956)

Chapter 5
Attractors in the Universe of Cellular Automata

One-dimensional cellular automata (CA) with $L = I + 1$ cells are *complex systems* with *nonlinear dynamics* (Alligood et al. 1996; Mainzer 2009; Shilnikov et al. 2001) determined by one of the 256 local rules N. Their *state spaces* contain all distinct states of cellular rows $\left(x_0^t, \ldots, x_{I-1}^t, x_I^t\right)$ at time step t (iteration or generation). An entire list of consecutive rows with no two rows identical and including the initial configuration is called an *orbit* in the state space of a cellular automaton. With that background, the well-known attractor dynamics of complex systems can also be studied in the theory of CA (Chua et al. 2005a).

Transient Regime and Basin of Attractors

For finite length I, the dynamic pattern evolving from the initial state $\mathbf{x}(0) = \mathbf{x}^0 = \left(x_0^0, \ldots, x_{I-1}^0, x_I^0\right)$ under any rule N must eventually repeat itself with a minimum period. The set of repeated cellular rows can be considered a *periodic attractor* which is characterized by its attractor period. The set of all initial states that tend to that attractor is called the *basin of attraction*. The first consecutive rows of the dynamic pattern from the initial state to the beginning of the periodic attractor are sometimes called the *transient regime*. The entire dynamic pattern is the orbit originating from the initial state.

In Fig. 1, attractor dynamics of CA are illustrated formally. The *transient duration* of the transient regime is denoted by T_δ. The *minimum period of repetition* in an attractor Λ is T_Λ. The *dynamic pattern* evolving from the initial state $\mathbf{x}(0)$ under any rule N is denoted by $D_N(\mathbf{x}(0))$. In Fig. 1a, the first dynamic pattern $D_{62}(\mathbf{x}_a(0))$ is shown for the initial configuration $\mathbf{x}_a$ in row 0 with $T_\delta = 51$ und $T_\Lambda = 3$. The transient regime originating from $\mathbf{x}_a$ of the dynamic pattern $D_{62}(\mathbf{x}_a(0))$ consists of the first 51 rows. The period-3 orbit is clearly seen by the alternating color of the background. For the dynamic pattern $D_{62}(\mathbf{x}_b(0))$, the initial configuration $\mathbf{x}_b$ in row 0 of Fig. 1b gives rise to a longer transient duration

K. Mainzer and L. Chua, *The Universe as Automaton*, SpringerBriefs in Complexity, DOI: 10.1007/978-3-642-23477-4_5, © The Author(s) 2012

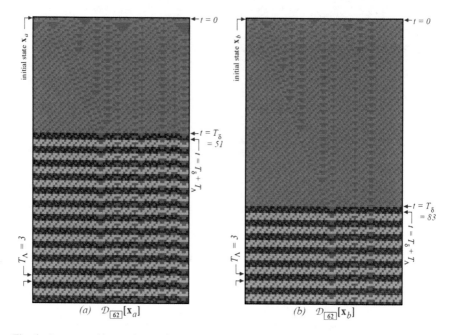

(a) $\mathcal{D}_{\boxed{62}}[\mathbf{x}_a]$ (b) $\mathcal{D}_{\boxed{62}}[\mathbf{x}_b]$

Fig. 1 Attractor with transient regime and transient duration of rule 62 originating from different initial configurations $\mathbf{x}_a$ and $\mathbf{x}_b$

$T_\delta = 83$. However, since $\mathbf{x}_a$ and $\mathbf{x}_b$ were chosen to belong to the basin of attraction of Λ, the period T_Λ of the periodic orbit in Fig. 1a, b must be the same, namely, $T_\Lambda = 3$.

There are rules such as 110 and their global equivalence classes, where T_Λ may tend to infinity as $T \to \infty$ for $I \to \infty$. In this case, instead of a transient regime, we refer to the entire dynamic pattern $D_N(\mathbf{x}(0))$ as an orbit originating from $\mathbf{x}(0)$. By definition the *basin of attraction* of an attractor Λ must contain at least one element, not belonging to Λ. But there are some periodic orbits that have no basin of attraction. They are said to be *invariant*.

Special configurations which have no predecessors with respect to rule N are called "*garden of Eden*" (Moore 1962). The intuitive association with the biblical garden of Eden is obvious. No garden of Eden can be a periodic orbit with a period more than 1, otherwise any point on the orbit is a predecessor of its next iterate. A truly unique species is an "*isle of Eden*": If the predecessor of a configuration with respect to rule N is itself, the configuration is said to be an "isle of Eden". We can distinguish *period-k isles of Eden* depending on k applications of rule N to generate itself. We will come back to these exotic phenomena in the universe of CA that have no counterpart in the theory of dynamical systems (Garay et al. 2008).

Characteristic Functions of Cellular Automata

Because each attractor of a cellular automaton is periodic (for finite I) with some period T_Λ, it is represented by consecutive bit strings $\mathbf{x}(0)$, $\mathbf{x}(1)$, $\mathbf{x}(2)$,..., $\mathbf{x}(T_A)$, as illustrated in Fig. 1. To apply the analytical tools of nonlinear dynamics, the *pattern formation of CA* must be transcribed into an equivalent *nonlinear time series*. In the following, we consider states in the *state space* of a cellular automaton N in the vector notation of a *Boolean string* $\overrightarrow{\mathbf{x}}(0) = [x_0(0), x_1(0),\ldots, x_{I-1}(0), x_I(0)]$ with $x_I(0) \in \{0, 1\}$ Each Boolean string $\overrightarrow{\mathbf{x}}$ can be associated uniquely with the binary expansion (in base 2) of a *real number* $\phi = 0 . x_0 x_1 \cdots x_{I-1} x_I$ on the unit interval $[0, 1]$. The *decimal equivalent* of ϕ is $\phi = \sum_{i=0}^{I} 2^{-(i+1)} x_i$. The bilateral image $\overleftarrow{\mathbf{x}}(0) = [x_I(0), x_{I-1}(0),\ldots, x_1(0), x_0(0)] = \mathbf{T}^\dagger(\overrightarrow{\mathbf{x}}(0))$ is called the *backward Boolean string* associated with the *forward Boolean string* $\overrightarrow{\mathbf{x}}$ by the $(I+1)$-dimensional *left–right transformation operator* $\mathbf{T}^\dagger$ (see Chap. 4). Each backward Boolean string $\overleftarrow{\mathbf{x}}$ maps into the real number $\phi^\dagger = 0 . x_I x_{I-1}\ldots x_1 x_0$, where the decimal equivalent is given by $\phi^\dagger = \sum_{i=0}^{I} 2^{-(i+1)+i} x_i$.

For a one-dimensional CA with $I + 1$ cells, there are $n_\Sigma = s^{I'}$ distinct Boolean strings with $I' = l + 1$. The *state space* Σ is the collection of all n_Σ Boolean strings. Each local rule N induces a *global map* $T_N : \Sigma \to \Sigma$, where each state $\mathbf{x} \in \Sigma$ is mapped into exactly one state $T_N(\mathbf{x}) \in \Sigma$. Since each state $\mathbf{x} \in \Sigma$ corresponds to one, and only one, point $\phi \in [0, 1]$, it follows that the global map T_N induces an equivalent map χ_N from the set of all rational numbers $\mathcal{R}[0, 1]$ over the unit interval $[0, 1]$ into itself, namely $\chi_N : \mathcal{R}[0, 1] \to \mathcal{R}[0, 1]$, called the *CA characteristic function* of N. In the limit where $I \to \infty$, the state space Σ coincides with the collection of all infinite strings extending from $-\infty$ to $+\infty$ and $\lim_{I\to\infty} \mathcal{R}[0, 1] = [0, 1.]$ In this general case, the CA characteristic function is defined at every point (or real number) $\phi \in [0, 1]$, including all irrational numbers (Mainzer 1990).

Since the domain of the CA characteristic function χ_N (for finite I) consists of a subset of rational numbers in the unit interval $[0, 1]$, a computer program for constructing the *graph of the characteristic function* χ_N can easily be written as follows (cf. Fig. 2):

Step 1: Divide the unit interval $[0, 1]$ into a finite number of uniformly-spaced points, called a *linear grid*, of width $\Delta\phi = 0.005$.

Step 2: For each grid point $\phi_j \in [0, 1]$, identify the corresponding binary string $s_j \in \Sigma$.

Step 3: Determine the image $s'_j \in \Sigma$ of s_j under rule N. In other words, find $s'_j = T_N(s_j)$ via the truth table of N.

Step 4: Calculate the decimal ϕ' equivalent of s'_j.

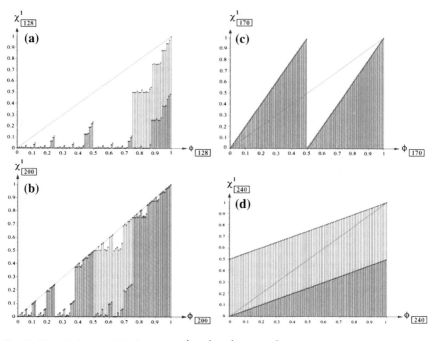

Fig. 2 Time-1 characteristic functions χ^1_{120}, χ^1_{200}, χ^1_{170}, and χ^1_{240}

Step 5: Plot a vertical line through the abscissa $\phi_N = \phi_j$ with height equal to s'_j.
Step 6: Repeat steps 1–5 over all $(1/\Delta\phi) + 1$ grid points. In Fig. 2, there are $(1/0.005) + 1 = 201$ grid points.

Sometimes, it is useful to plot the τth iterated value $s^\tau_j = T^\tau_N(s_j) = T_N \circ T_N \circ \cdots \circ T_N(s_j)$ of s_j, instead of $T_N(s_j)$ at each grid point $\phi_j \in [0, 1]$. Such a function is called a *time*−τ *CA characteristic function* which is denoted with χ^τ_N. The introduced algorithm can be used to plot the graph of the time-1 CA characteristic function χ^1_N of any rule N. The same algorithm can be used for plotting the graph of the time−τ CA characteristic function χ^τ_N as well.

To visualize the complex patterns of Fig. 2, the vertical lines of χ^1_N are plotted in alternating red and blue colors, referred to as red and blue-coordinates ϕ_{red} and ϕ_{blue}. All members of the red group have a 0 as their rightmost bit. The blue group then consists of binary strings with a 1 as their rightmost bit. Since the rightmost (end) bit of each $\phi_{blue} \in [0, 1]$ is equal to a 1, it follows that the largest value of ϕ_{blue} is greater than the largest value of ϕ_{red} by exactly $1/2^{I+1}$. Therefore, the rightmost vertical line must be blue in color, and tends to $\phi = 1$ as $I \rightarrow \infty$ (For plotting purposes, the rightmost blue line is drawn through $\phi = 1$.) The 201 red and blue lines shown in Fig. 2 represent only their approximate positions on $[0, 1]$, because the resolution of their exact positions is determined by the value of I, which is chosen to be 65 in Fig. 2. Thus, the *state space* Σ is coarse-grained and only contains 2^{66} distinct 66-bit binary strings, each one representing a unique rational number on [0, 1].

The graph of the characteristic function χ_{128}^1 of rule 128 in Fig. 2a is very simple. No vertical line intersects the unit-slope main diagonal except at $\phi_{128} = 0.000\ldots$ and $\phi_{128} = 1.000\ldots$ These two period-1 fixed points give rise to a homogeneous pattern $D_{128}(0.000\ldots)$ of blue color (value 0) and a homogeneous pattern $D_{128}(1.000\ldots)$ of red color (value 1). But these two orbits have different kinds of dynamics. The orbit from $D_{128}(0.000\ldots)$ is an *attractor* in the sense of nonlinear dynamics, because it has a nonempty basin of attraction, which consists of all points in the closed-open interval [0,1).

The graph of the characteristic function χ_{200}^1 of rule 200 in Fig. 2b has many vertical lines terminating exactly on the main diagonal. They indicate many *period-1 fixed points* with many *period-1 attractors*.

The graph of the characteristic function χ_{170}^1 of rule 170 in Fig. 2c has no period-1 fixed points except at $\phi_{170} = 0.000\ldots$ and $\phi_{170} = 1.000\ldots$. The vertices of all vertical lines fall on one of two parallel lines with slope 2. This is an example of a so-called *Bernoulli shift*.

The graph of the characteristic function χ_{240}^1 of rule 240 in Fig. 2d also has no period-1 fixed points except at $\phi_{240} = 0.000\ldots$ and $\phi_{240} = 1.000\ldots$. All red vertical lines terminate on the lower straight lines of slope = ½, and all blue vertical lines terminate on the upper parallel straight lines. The two piecewise-linear functions χ_{240}^1 and χ_{170}^1 are the *inverse* of each other. χ_{240}^1 is also an example of a *Bernoulli-shift*.

In general, each CA local rule N can exhibit *many distinct attractors* Λ_i. Each attractor represents a distinct pattern and must be analyzed as a separate dynamical system. The *left–right transformation operator* $\mathbf{T}^\dagger$ allows us to study the *lateral symmetry* of bilateral pairs N and $N^\dagger = \mathbf{T}^\dagger(N)$ for local rules. It is useful, therefore, to consider each attractor from two spatial directions, namely, a forward (left $\to$ right) direction and a backward (right $\to$ left) direction. Each period-T_Λ attractor Λ, defined by a pattern of T_Λ consecutive Boolean strings, can be mapped onto a *forward time series* $\varphi = \left[\phi_0, \phi_1, \ldots, \phi_{T_\Lambda}\right]$ with $\phi_i \in [0, 1]$, called a *forward orbit*, and a *backward time series* $\varphi^\dagger = \left[\phi_0^\dagger, \phi_1^\dagger, \ldots, \phi_{T_\Lambda}^\dagger\right]$ with $\phi_i^\dagger \in [0, 1]$, called a *backward orbit*, with length T_Λ for each time series.

The dynamics of an attractor can be illustrated and understood by plotting the two attractor-induced time-τ maps (Alligood et al. 1996), associated with the forward time series φ and the backward time series $\varphi^\dagger$. For each rule N, the *forward* time-τ map $\rho_\tau : \phi_{n-\tau} \mapsto \phi_n$ is defined by the time-τ *characteristic function* χ_N^τ with $\rho_\tau(\phi_{n-\tau}) = \chi_N^\tau(\phi_{n-\tau})$, and the *backward* time-τ *map* $\rho_\tau^\dagger : \phi_{n-\tau}^\dagger \mapsto \phi_n^\dagger$ is defined by the time-τ *characteristic function* χ_N^τ with $\rho_\tau^\dagger\left(\phi_{n-\tau}^\dagger\right) = \chi_N^\tau\left(\phi_{n-\tau}^\dagger\right)$.

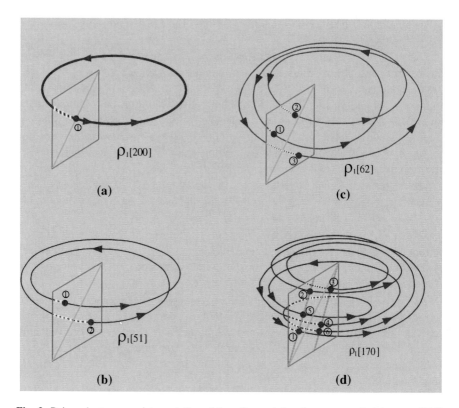

Fig. 3 Poincaré return map interpretation of three forward time-1 maps: **a** period-1 map $\rho_1[200]$, **b** period-2 map $\rho_1[51]$, **c** period-3 map $\rho_1[62]$, **d** Bernoulli shift map $\rho_1[170]$

Poincaré Return Maps of Cellular Automata

When $\tau = 1$, the time-1 maps ρ_τ and $\rho_\tau^\dagger$ (Alligood et al. 1996; Hirsch et al. 1974) behave like *Poincaré return maps* (Poincaré 1897) in the theory of dynamical systems. In Fig. 3, the three forward time-1 maps $\rho_1[200], \rho_1[51]$, and $\rho_1[62]$ of rules $N = 200, 51$, and 62 are illustrated as Poincaré return maps with a Poincaré cross-section in the unit-square $[0, 1] \times [0, 1]$. In Fig. 3a, only one *period-1 attractor* of rule 200 is labeled as point 1. All iterates from points inside the basin of attraction map onto the fixed point 1. One can imagine a planet intersecting an imaginary Poincaré cross-section once every revolution.

Figure 3b shows a *period-2 orbit* (isle of Eden) of local rule 51. The orbit of the circulating planet intersects the Poincaré cross-section at two points with $\rho_1(1) \mapsto 2$ and $\rho_1(2) \mapsto 1$. Figure 3c shows a *period-3 attractor* of local rule 62. The circulating orbit intersects the Poincaré cross-section at three points with $\rho_1(1) \mapsto 2, \rho_1(2) \mapsto 3$, and $\rho_1(3) \mapsto 1$.

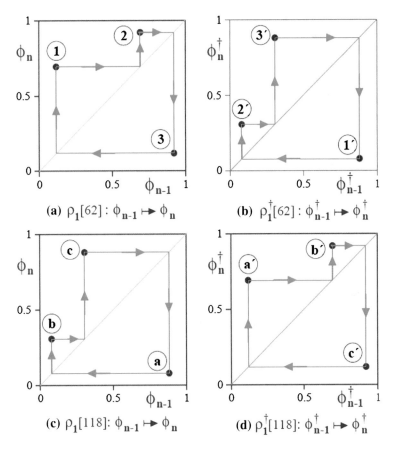

(a) $\rho_1[62] : \phi_{n-1} \mapsto \phi_n$

(b) $\rho_1^\dagger[62] : \phi_{n-1}^\dagger \mapsto \phi_n^\dagger$

(c) $\rho_1[118] : \phi_{n-1} \mapsto \phi_n$

(d) $\rho_1^\dagger[118] : \phi_{n-1}^\dagger \mapsto \phi_n^\dagger$

Fig. 4 Cobweb diagram with the evolution of CA rules $N = 62$ and 118

Figure 3d shows a *Bernoulli-shift orbit* (isle of Eden) of rule 170 (compare Fig. 2c). The circulating orbit intersects the Poincaré cross-section at almost all points on the two parallel lines with slope equal to 2 for the case $I = \infty$. Only a few iterates 1, 2, 3, 4, 5, 6 are shown. The orbit reminds us of a comet visiting almost all points on the two parallel lines when I goes to infinity.

Lameray Diagrams of Cellular Automata

Cellular Automata rules of the same *global equivalence class* have identical behaviors. In particular, they have the same transient regimes, the same attractors, and the same invariant orbits, with respect to a bijective mapping. For example, let us consider rules 62 and 118, belonging to the same global equivalence class ε_{22}^2. They are related by a left–right transformation operator $118 = \mathbf{T}^\dagger(62)$. In Fig. 4, their forward and backward time-1 maps (a and c) and backward time-1 maps (b and d) are illustrated in a so-called *Lameray diagram* (Shilnikov et al. 1998), named after the French mathematician Lameray who used its intuitive value back in the eighteenth century. It is also called a *cobweb diagram* (Alligood et al. 1996) because it resembles the web spun by a spider.

A cobweb plot is a visual tool used in dynamical systems to investigate the qualitative behavior of one-dimensional iterated functions. Using a cobweb plot, it is possible to infer the long-term status of an initial condition under repeated application of a map. For a given iterated function $\rho_1[N] : \phi_{n-1} \mapsto \phi_n$ of CA rule N, the plot consists of a diagonal line with $\phi_{n-1} = \phi_n$ and a curve representing $\phi_n = \rho_1[N](\phi_{n-1})$. To plot the behavior of a value ϕ_0, apply the following steps:

Step 1: Find the point on the function curve with an ϕ_{n-1}-coordinate of ϕ_0. This has the coordinates $(\phi_0, \rho_1[N](\phi_0))$.
Step 2: Draw a horizontal projection line from this point to the diagonal line. This has the coordinates $(\rho_1[N](\phi_0), \rho_1[N](\phi_0))$.
Step 3: Draw a vertical projection line from the point on the diagonal to the function curve. This has the coordinates $(\rho_1[N](\phi_0), \rho_1[N](\rho_1[N](\phi_0)))$.
Step 4: Repeat from step 2 as required.

In Fig. 4, the cobweb diagrams show the evolution of CA rules 62 and 118 from any state of a *period-3 attractor in forward time* (a and c) and *backward time* (b and d). In Fig. 4a, 1, 2, and 3 are the three points of the circulating orbit of $\rho_1[62] : \phi_{n-1} \mapsto \phi_n$. In Fig. 4b, 1′, 2′, and 3′ are the three points of the circulating orbit of $\rho_1^\dagger[62] : \phi_{n-1}^\dagger \mapsto \phi_n^\dagger$. In Fig. 4c, a, b, and c are the three points of the circulating orbit of $\rho_1[118] : \phi_{n-1} \mapsto \phi_n$, and in Fig. 4d, a′, b′ and c′ are the three points of the circulating orbit of $\rho_1^\dagger[118] : \phi_{n-1}^\dagger \mapsto \phi_n^\dagger$.

The dynamics and long-term behavior of each attractor of a local rule N can often be predicted from one or more of its time-τ maps $\rho_\tau[N]$. Many rules have attractors similar to those in Fig. 3. For *forward time-1 maps* $\rho_\tau[N]$ and *backward time-1 maps* $\rho_\tau^\dagger[N]$, distinct attractors can be identified by *cobweb diagrams* where points of different colors belong to different attractors.

Power Spectrum of Cellular Automata

In Fig. 5, the forward time-1 map is also characterized by the *power spectrum* of the associated forward time series $\varphi = [\phi_0, \phi_1, \ldots, \phi_{T_\Lambda}]$ with $\phi_i \in [0, 1]$. In the theory of dynamical systems, the power spectrum is a useful instrument in analyzing additional information that is not revealed from time-τ maps. The cellular automaton dynamics of CA calculated from the total of each step of the associated time series, may be seen as a kind of signal and analyzed by its power spectrum. In signal theory, the power spectrum found by the discrete Fourier transform of CA dynamics shows the amount of energy over time and illustrates *periodicities*

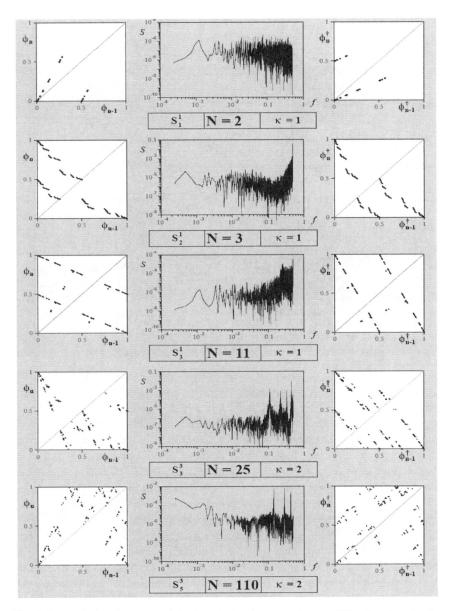

Fig. 5 Forward time-1 maps and backward time-1 maps of some CA rules for attractor $\Lambda_1(red), \Lambda_2(blue)$, and $\Lambda_2(green)$ with corresponding power spectra

(Kaplan et al. 1995). It describes how the power of a signal or time series is distributed with frequency. Thus, the power spectrum of all periodic-1 time-1 maps consists of one line only emerging at *frequency* $f = 1$ signifying the absence of any other frequency components. The power spectrum of all period-2 time-1

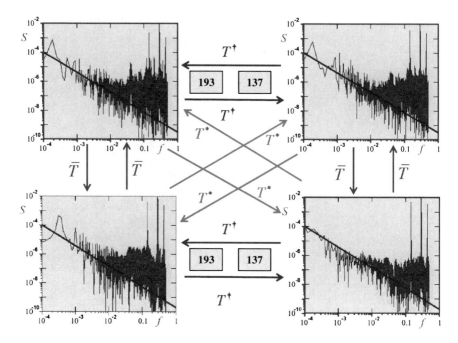

Fig. 6 The power spectrum of the four globally equivalent rules 110, 124, 137, and 193 capable of universal computation with 1/f power frequency

maps consists of a line located at $f = \frac{1}{2}$. In Fig. 5, rule $N = 2$ with *complexity degree* $\kappa = 1$ shows the time-1 map (only in red) of just one attractor. In Fig. 5, rule 25 with complexity degree $\kappa = 2$ shows three time-1 maps (colored in red, blue, and green) corresponding to three distinct types of attractors.

A careful analysis of the power spectra of all 256 rules reveals a remarkable property of Klein's *symmetry group* associated with the four globally-equivalent rules 110, 124, 137, and 193. The "*holy grail*" in the universe of CA does not only seem to be distinguished by universal computability, but also by a *1/f—power frequency* with a slope equal to approximately -1.5 (Schröder 1991). This observation suggests that there might exist a fundamental relationship between *universal computation* and the *1/f—power laws* (Fig. 6). Actually the determination of their low-frequency spectra requires an immense amount of simulation time. In general, in complex dynamical systems, power laws indicate a *high degree of complexity* (Mainzer 2007a). In natural sciences and economics, they are correlated with extreme events such as tsunamis, financial crises, or technical disasters (Albeverio et al. 2006). Thus, the CA of Klein's symmetry group with universal computability may be assumed to simulate complex processes in nature, society, and technology. These insights are absolutely amazing with regard to the simplicity of the local rules generating the complex global evolution of the CA 110, 124, 137, and 193.

Invertible Attractors

How can the *long-term behavior* of complex pattern formation be *predicted*? An accurate analysis of certain rules leads to remarkable insights. The forward time-1 maps and backward time-1 maps in Fig. 5 reveal *symmetries* with respect to the main diagonal in the left and right frames. Since the period T_Λ of any attractor Λ is the smallest integer where the orbit repeats itself, no two points in the domain of the functions $\rho_\tau[N]$ and $\rho_\tau^\dagger[N]$ can map to the same point. Therefore, both maps $\rho_\tau[N]$ and $\rho_\tau^\dagger[N]$ are bijective, and hence have a well-defined single-valued *inverse* map, $[\rho_\tau[N]]^{-1}$ and $\left[\rho_\tau^\dagger[N]\right]^{-1}$, respectively. The *forward* time-τ map $\rho_\tau[N]$: $[0, 1] \rightarrow [0, 1]$ is said to be *invertible* over the interval $[0, 1]$ iff $\rho_\tau[N] = \left[\rho_\tau^\dagger[N]\right]^{-1}$. In this case, the map is symmetrical with respect to the main diagonal. The *backward* time-τ map $\rho_\tau^\dagger[N]$: $[0, 1] \rightarrow [0, 1]$ is said to be *invertible* over the interval $[0, 1]$ iff $\rho_\tau^\dagger[N] = [\rho_\tau[N]]^{-1}$. In this case, the map is also symmetrical with respect to the main diagonal (Chua et al. 2005a; Toffoli et al. 1990).

The sets of points in the left and right frames in Fig. 5 are called the graphs of the time-1 maps $\rho_1[N]$ and $\rho_1^\dagger[N]$. The left and right frames of rule 3 have only one color (red). Thus, the corresponding cellular automaton has only one robust attractor. (Robust attractor means roughly that there is a sufficiently large basin of attraction for there to be a good chance that an arbitrarily chosen initial state will converge to it.) Since the graph of $\rho_1[3]$ on the left and the graph of $\rho_1^\dagger[3]$ on the right are reflections of each other about the main diagonal, the time-1 maps $\rho_1[3]$ and $\rho_1^\dagger[3]$ are *invertible*. The two colors in the left and right frames of rule 11 correspond to two robust attractors. Since both graphs of the same color are mirror images about the diagonal, both pairs of time-1 maps of 11 are *invertible*. In the case of rule 110, the red color graphs on the left and right sides are not mirror images of each other. Therefore, the forward time-1 map $\rho_1[110]$ and backward time-1 map $\rho_1^\dagger[110]$ are *not invertible*.

The graphs of both forward and backward time-1 maps of all 256 CA rules provide deep insights into the dynamics of CA. Because these graphs do not depend on the initial state, they completely characterize the long-term behavior of all rules, and allow one to *predict* their *long-term development*. Each CA rule can have several attractors and invariant orbits. A CA rule N is either *bilateral* when $N^\dagger = T^\dagger(N)$, or *nonbilateral*. It can be either *invertible* when its forward and backward time-1 maps are symmetrical with respect to the main diagonal, or *noninvertible*.

Finally, we list some results without further proofs (Chua et al. 2005a): There are 45 *invertible* and 24 *noninvertible period-1 rules*. Each period-1 rule generally has a continuum of period-1 attractors, clustered along the main diagonal. Among

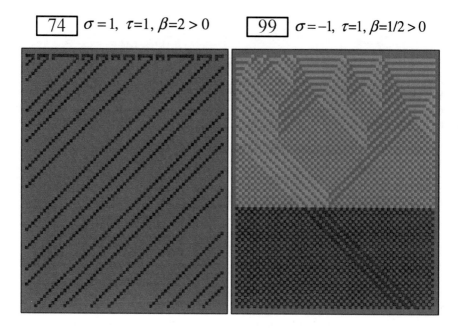

Fig. 7 Dynamic patterns of Bernoulli rules 74 (**a**) and 99 (**b**)

the period-1 rules, there are 12 rules which always tend to the homogeneous attractor with state 0, and another 12 rules which always tend to the homogeneous attractor with state 1, independent of the initial state (except for the isles of Eden states, which we will consider later on). There are 17 *invertible period-2 rules* all of which are *bilateral*. There are also *eight noninvertible period-2 rules*, all of which are *nonbilateral*. Period-2 attractors are manifested by a symmetrical pair of points lying on an imaginary line drawn perpendicular to the main diagonal. In addition, there are four *noninvertible period-3 rules*. Thus, we have so far listed $45 + 24 + 17 + 8 + 4 = 98$ rules, the long-term behavior of which can be predicted with respect to the properties mentioned.

Bernoulli Shifts of Cellular Automata

In addition to the rules mentioned in the previous chapter, there are 108 rules whose attractors can be precisely *predicted* by so-called *Bernoulli shift maps,* which have already been introduced with rules 170 and 240 (Fig. 2c, d). The evolution of each initial configuration of these 108 rules can be predicted by shifting it either to the left, or to the right, by 1, 2, or 3 pixels, possibly followed by a complementation, which means a change of color.

For example, consider the *dynamic patterns* $D_n(\mathbf{x}(0))$ of the Bernoulli rules $N = 74$ and 99 in Fig. 7a, b. Let $\mathbf{s}^t$ be any row on the attractor of these two patterns. Formally we choose $t > T_\delta$ where T_δ denotes the *transient duration* of the *transient regime* to the *attractor*. Obviously the evolution rule for CA 74 demands: Shift string $\mathbf{s}'$ to the left by one pixel to obtain the first iteration $\mathbf{s}^{t+1}$. Repeating the same procedure we obtain the same pattern shown in Fig. 7a. Similarly, the rule for 99 demands: Shift string $\mathbf{s}'$ to the right by one pixel to obtain the first iteration $\mathbf{s}^{t+1}$. Repeating the same procedure we obtain the periodic pattern with period $T = I + 1$ after the transient regime shown in the upper part of Fig. 7b, where the high-lit area denotes the attractor regime.

In general, each Bernoulli shift is uniquely identified by *three parameters* σ, τ, and β. σ is a positive or negative integer. A positive σ means "shift the bit string by σ bits to the left". A negative σ means "shift the bit string by $|\sigma|$ bits to the right". τ is a positive integer, indicating the shifted bit string from the previous operation is not the next bit string, but the bit string τ rows below it. β is either a positive or a negative number. A positive β requires no change. A negative β requires taking the complement (i.e., change 0 to 1, and 1 to 0) of the shifted bit string. Among the 108 *Bernoulli σ_τ-shift rules*, there are 80 invertible rules, organized as members of 24 global equivalence classes ε_m^k with different *Bernoulli-attractors*. There are 28 noninvertible Bernoulli σ_τ-shift rules with Bernoulli attractors.

Bernoulli shift rules reveal the hidden secrets of the corresponding CA dynamics in their *binary coding*. For more details, let us consider the shifting rule ($\sigma = -1 < 0$) of CA 240 with *characteristic function* χ_{240}^1 in Fig. 2d. The first few digits of the decimal expansion of a binary bit string $\overrightarrow{\mathbf{x}} = [x_0, x_1, \ldots, x_{I-1}, x_I]$ with $x_i \in \{0, 1\}$ is given by $\phi = \frac{1}{2}x_0 + \frac{1}{4}x_1 + \frac{1}{8}x_2 + \cdots + \frac{1}{2^I}x_{I-1} + \frac{1}{2^{I+1}}x_I = 0.5x_0 + 0.25x_1 + 0.125x_2 + \cdots + \frac{1}{2^I}x_{I-1} + \frac{1}{2^{I+1}}x_I$.

If the left-most bit is $x_0 = 0$, then $\phi < 0.5$ and the lower straight line with slope $\frac{1}{2}$ in Fig. 2d will be selected. If the leftmost bit is $x_0 = 1$, then $\phi > 0.5$ and the upper branch in Fig. 2d will be selected. If the rightmost (end) bit is $x_I' = 1$ at time t, then the *Bernoulli σ_1-shifting rule* of CA 240 shifts the end bit 1 to the right. Thereby, according to the periodic boundary condition in Fig. 1a of Chap. 2, it will reappear as the first bit in the next iteration. Since the first bit in the next iteration reads $x_0^{t+1} = 1$, we have $\phi^{t+1} > 0.5$ and the dynamics follow the *upper branch* of χ_{240}^1. Conversely, if the end bit is $x_I' = 0$, then the Bernoulli right shifting rule for CA 240 shifts the bit 0 to make the first bit equal to $x_0^{t+1} = 0$ in the next iteration. In this case, the dynamics follow the *lower branch* of χ_{240}^1.

The right shifting rule of CA 240 can be illustrated in a *cobweb diagram* (Fig. 8) showing a succession of ten iteration points 1, 2, 3, ..., 9, 10 undergoing the σ_1-shifting evolution dynamics. The decimal coordinate $\phi_0 = 0.673768048097057\ldots$ of point 1 is calculated from the following 66-bit string (with $I = 65$ in Fig. 2) with the conversion formula that was introduced above:

101011000111110000010000000100111010101010100111010110010101110101

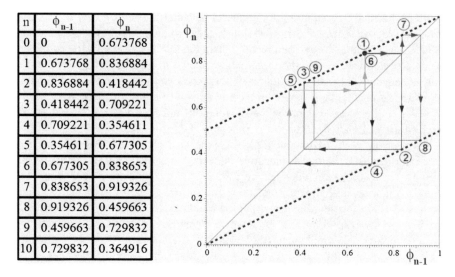

n	ϕ_{n-1}	ϕ_n
0	0	0.673768
1	0.673768	0.836884
2	0.836884	0.418442
3	0.418442	0.709221
4	0.709221	0.354611
5	0.354611	0.677305
6	0.677305	0.838653
7	0.838653	0.919326
8	0.919326	0.459663
9	0.459663	0.729832
10	0.729832	0.364916

Fig. 8 Cobweb diagram of Bernoulli rule 240 with σ_1-shifting evolution dynamics

The rightmost (end) bit of the above bit string is a 1. To obtain the next iteration via the σ_1-*shifting rule* for $N = 240$, we simply shift the above bit string by *one pixel* (since $\tau = 1$) to the right (since $\sigma < 0$). Again, with respect to the periodic boundary condition of Fig. 1a of Chap. 2, 1 is inserted at the leftmost (first) position of the above right-shifted string to obtain the following 66-bit string

1 1010110001111100000100000001001110101010101001110101100101011010

with the decimal value $\phi_1 = 0.836884024048529....$ Leibniz would be delighted that while the two decimal numbers ϕ_0 and ϕ_1 seem to be without any relation, their binary codes reveal a trivial *Bernoulli right shift of one pixel*.

Bernoulli Shifts and Coin-Toss Experiments

In Fig. 8, the Bernoulli shift rule seems to converge to two parallel lines. But this is only true for the limiting case $I \to \infty$. For finite I, it is an illusion, caused by the poor resolution of a printer or of our retina. Actually, there are tiny differences in different *parallel lines depending* on the finite size of I. In Fig. 8, points 1 and 6 appear as a single point because they differ by only 0.003537. But the *cobweb diagram* starting from 6 instead of 1 would evolve into an entirely different orbit. This observation reveals the well-known *extreme sensitivity of nonlinear dynamics*. In general, the Bernoulli-shift for $I \to \infty$ is as *chaotic* as a coin toss (Nagashima et al. 1999). Its chaotic attractor has a *Lyapunov exponent* $\lambda = \beta = 2 > 1$ (Devaney 1992). Intuitively, the Lyapunov exponent reflects the concept of weak causality. According to *weak causality*, in unstable and chaotic situations, tiny and local causes can lead to large and global effects, contrary to *strong causality*, when similar (tiny or large) causes lead to similar (tiny or large) effects. Mathematically, the Lyapunov exponent of a dynamical system is a quantity that characterizes the rate of separation of infinitesimally close trajectories. Two trajectories in the phase space of the dynamical system with a certain initial separation diverge according to a measure depending on an exponent λ, called the Lyapunov exponent. The rate of separation can be different for different orientations of the initial separation vector. Thus, there is a spectrum of Lyapunov exponents, which is equal in number to the dimensionality of the phase space. The largest (maximal) Lyapunov exponent determines a notion of predictability for a dynamical system. A positive maximal Lyapunov exponent is usually taken as an indication that the system is *chaotic* provided some other conditions are satisfied (Kaplan et al. 1995; Mainzer 2007b).

For more details about *coin-toss experiments* and Bernoulli shifts let us consider the inverse Bernoulli rule 170, which has identical dynamics to rule 240, as $I \rightarrow \infty$. There is a one-to-one correspondence between the iterations of the Bernoulli rule 170 and the outcome of an ideal coin-toss. In Fig. 2, the characteristic function $\chi_{170}^1 : [0,1] \rightarrow [0,1]$ can be described analytically by $\chi_{170}^1 = 2\phi_{170}$ mod 1 for all $\phi_{170} \in [0,1]$. Therefore, every point of the unit interval $[0,1]$ corresponds to a semi-infinite binary bit string $[x_0, x_1, \ldots, x_{I-1}, x_I] \mapsto 0 \cdot x_0 x_1 \cdots x_{I-1} x_I$ where $I \rightarrow \infty$. Nearly all points in the interval $(0,1)$ (except those representing the rational numbers) correspond to an irrational number, whose binary expansion can be identified with a particular coin-toss experiment. The ensemble of all possible ideal coin-toss experiments corresponds to the set of all points on $[0,1]$. To illustrate any coin-toss experiment, we choose an arbitrary point from the unit interval $[0,1]$. Then, with the Bernoulli left-shifting rule $\sigma_1[170]$, we can read out the first digit x_0^{t+n} from each iteration $t + n$. The outcome of this binary output string is also a member of the ideal coin-toss ensemble. In this sense, the Bernoulli rule 170 and its inverse rule 240 are as *chaotic* as an ideal coin-toss, as $I \rightarrow \infty$.

Fractality of Cellular Automata

So far we have completely characterized the long-term behavior of $108 + 98 = 206$ one-dimensional CA rules with three inputs. 108 CA can be predicted by *Bernoulli-shift rules*. 98 CA are characterized as 45 invertible and 24 noninvertible *period-1 rules*, 17 *invertible period-2 rules* all of which are *bilateral*, eight *noninvertible period-2 rules* all of which are *nonbilateral*, and four *nonbilateral period-3 rules*, which can be either invertible or noninvertible. The remaining 50 rules of the 256 CA rules consist of 18 *noninvertible* but *bilateral* (called "*complex*" Bernoulli rules), and 32 *noninvertible* and *nonbilateral* rules (called "*hyper*" Bernoulli rules). These 50 remaining rules can be reduced to 18 global equivalence

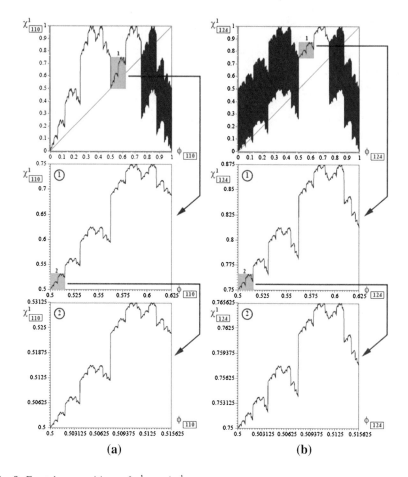

Fig. 9 Fractal compositions of χ^1_{110} and χ^1_{124}

classes. It is sufficient to consider only one representative from each equivalence class because of the similar dynamics of their elements. The qualitative long-term dynamics of the 50 remaining rules can be studied with respect to further criteria of attractor dynamics.

A well-known property of attractor dynamics is *fractality*. A careful examination of the characteristic functions χ^1_N reveal that nearly every graph of χ^1_N exhibit a *fractal geometry* where *self-similar* two-dimensional substructures manifest themselves, going to infinity, as the number of cells $(I + 1) \rightarrow \infty$. In this case, subpatterns can be rescaled by appropriate horizontal and vertical scaling factors so that it coincides with a part of the composite patterns. The graphs of the characteristic functions χ^1_{110} and χ^1_{124} are plotted in Fig. 9. By rescaling the rectangle 1 by appropriate scaling factors, we obtain the corresponding subpattern 2. They are identical. In Fig. 9a, subpattern 1 of χ^1_{110} has the horizontal scaling $= 2^3$

and vertical scaling $= 2^2$, subpattern 2 of χ_{110}^1 has the horizontal scaling $= 2^6$ and vertical scaling $= 2^5$. In Fig. 9b, subpattern 1 of χ_{124}^1 has the horizontal scaling $= 2^3$ and vertical scaling $= 2^3$, subpattern 2 of χ_{124}^1 has the horizontal scaling $= 2^6$ and vertical scaling $= 2^6$. Continuing this process, we found the graphs of χ_{110}^1 and χ_{124}^1 are composed of infinitely many scaled self-similar patterns. Obviously, characteristic functions of this kind and the corresponding CA can be analyzed and *predicted* using properties of *fractal geometry*.

The graphs of the characteristic function χ_{170}^1 and χ_{240}^1 (Fig. 2c, d) do not have this kind of fractality. Their global dynamics are described by (mod 1) functions, as explained above. There are characteristic functions for eight rules that do not share the self-similarity of fractal geometry.

Gardens of Eden and Isles of Eden

At the beginning of this chapter, we mentioned the extraordinary phenomena of "gardens of Eden" and "isles of Eden". How can they be *analytically* defined by *characteristic functions* χ_N^1? *Gardens of Edens* are configurations of a cellular automaton with rule N which have no predecessors with respect to rule N. In more precise words, an $(I + 1)$-bit binary string $[x_0, x_1, \ldots, x_{I-1}, x_I]$ is said to be a garden of Eden of a CA rule N iff it does not have a predecessor under the corresponding transformation T_N. It follows that a garden of Eden $\phi_0 = \sum_{i=0}^{I} 2^{-(i+1)} x_i$ of N can never occur as a point on an orbit of N arising from some initial bit-string configuration whose decimal equivalent is different from ϕ_0. Thus, a garden of Eden has *no past*, but only *present* and *future* (Moore 1962).

Any binary string $[x_0, x_1, \ldots, x_{I-1}, x_I] \mapsto \phi_0 = \sum_{i=0}^{I} 2^{-(i+1)} x_i$ which has no preimage under χ_N^1 is a garden of Eden of rule N. It follows that $\phi_0 \in [0, 1]$ is a garden of Eden of N if a $\phi_{-1} \in [0, 1]$ does not exist such that $\phi_0 = \chi_N^1(\phi_{-1})$, where $\phi_{-1} \neq \phi_0$. But this property is only a sufficient condition for ϕ_0 to be a garden of Eden. There are special points that violate this property, but which are nevertheless gardens of Eden because they satisfy the definition. There exist special period-1 points which have no predecessors in the sense that no orbits from other initial bit-string configurations can converge to such points. According to the definition, they are also gardens of Eden, but as period-1 points they have a preimage under χ_N^1, namely, themselves. They are called *isles of Eden* (Chua et al. 2005b). In a literary sense, they have no past (like gardens of Eden), but also no future.

The concept of a *period-1 isle of Eden* can be generalized for time-n characteristic functions χ_N^n. Since such special period-n points also have no predecessors under the nth iterated map χ_N^n, they are called *period-n isles of Eden*. In other words, a bit string $[x_0, x_1, \ldots, x_{I-1}, x_I] \mapsto \phi_0 = \sum_{i=0}^{I} 2^{-(i+1)} x_i$ is said to be a period-n isle of Eden of rule N iff its preimage under χ_N^n is itself. In short, ϕ_0 is a

Fig. 10 Firing patterns of
rule 62

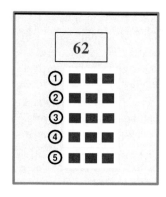

period-n isle of Eden of rule N iff $\chi_N^{-n}(\phi_0) = \phi_0$. It follows that a bit string ϕ_0 is a period-n isle of Eden of N iff ϕ_0 belongs to a period-n orbit with an empty basin tree. Observe that since no bit strings from a "period-n" isle of Eden with $n > 1$ qualifies as a garden of Eden, the two concepts of "Garden of Eden" and "Isle of Eden" are different when the period n of an isle of Eden is greater than 1.

We will come back to these particular phenomena of CA later on. They can only be found by careful computer search. From the analytical point of view in dynamical systems, they are surprising phenomena, because they have no counterpart in differential equations (Garay and Chua 2008).

Basin Trees of Attractors

To understand attractors, orbits, gardens of Eden, and isles of Eden more clearly, it is convenient to illustrate the basins of attraction by so-called *basin trees*. For example, we consider the cellular automaton with rule $N = 62$ which has the above five "*firing patterns*" determining a red cell with 1 (Fig. 10). The other three patterns are "non firing" determining a blue cell with state 0.

Under these rules, any bit string made of all zeros (blue bits) is a non firing pattern of 62. It can also be proven analytically by calculation that the corresponding number $\phi = 0$ is a fixed point of the characteristic function χ_{62}^1 of CA 62. We remember that the set of all string configurations converging to an attractor $\Lambda(62)$ of CA 62 is called the *basin of attraction* of $\Lambda(62)$. A basin of attraction can be illustrated in the form of a tree with nodes representing the string configurations of the basin and with edges indicating their application with rule 62. The numbers in the nodes represent the binary numbers of the corresponding string configuration. Figure 11 shows the *basin trees* of the period-1 point $\phi = 0$ of rule 62 for $I + 1 = 3$ and 4. The three or two pink bit strings, respectively, on the outer periphery in Fig. 11a, b of these basin trees are *gardens of Eden*, because they have no predecessors. The *self loop* at $\phi = 0$ means the repetitions of a fixed point.

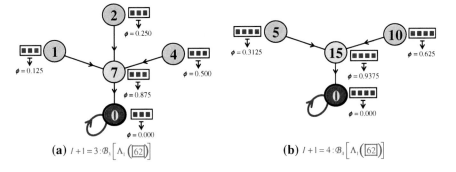

(a) $l+1=3:\mathcal{B}_3\left[\Lambda_1\left(\boxed{62}\right)\right]$ **(b)** $l+1=4:\mathcal{B}_3\left[\Lambda_1\left(\boxed{62}\right)\right]$

Fig. 11 Basin trees of the period-1 attractor $\phi=0$ for $l+1=3$ **(a)** and $l+3=4$ **(b)**

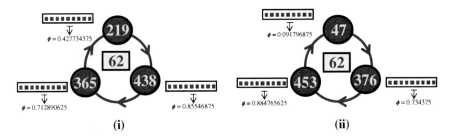

Fig. 12 State transition diagrams of isles of Eden with nine-bit strings

Figure 12 shows two period-3 orbits as *isles of Eden,* because they have an empty basin of attraction. None of these three-member triads has a predecessor except themselves. Such period-3 isles of Eden can be represented by an isolated single-loop state transition diagram.

Figure 13 shows the *basin tree of a period-3 attractor* with 6-bit strings. They are sometimes very complex and can be found systematically by brute force computer simulations.

There are also *Bernoulli* σ_τ-*shift attractors.* In the case of $\sigma=-1$ and $\tau=2$, the dynamics on the attractor consist of shifting each bit string on the attractor one bit to the right every two iterations. Figure 14a shows the *basin tree* of a corresponding example. Figure 14b shows a period-3 *isle of Eden* with an empty basin of attraction. Note that none of the bit strings from this isle of Eden is a *Garden of Eden.*

Although *"isles of Eden"* are non-robust orbits, almost all (in particular, 228 out of 256) rules harbor some isles of Eden, and some with *very large periods.* For example, Wolfram's celebrated "random-number generator" rule 30 has a 27-bit period-3240 isle of Eden. Since this is the *longest-period isle of Eden* known to date, the reader is encouraged to generate this rare "gem" by iterating the following 27-bit string 000000000000010010011101101 and verifying that it is indeed an isle of Eden, with a period equal to 3240 (Chua et al. 2008, Fig. 9,

Fig. 13 Basin tree of a
period-3 attractor with six-bit
strings

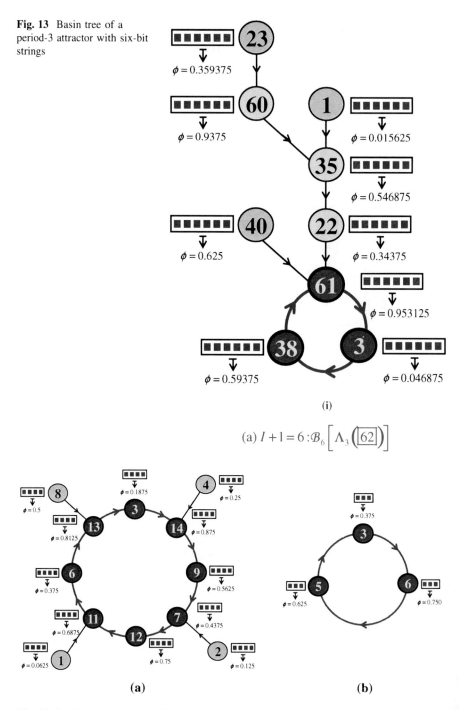

(i)

(a) $I + 1 = 6 : \mathcal{B}_6 \left[\Lambda_3 \left(\boxed{62} \right) \right]$

(a)   (b)

Fig. 14 Basin tree of a Bernoulli σ_τ-shift attractor with basin tree (**a**) for $I = 4$, and an isle of
Eden (**b**) for $I = 3$. Both periodic orbits are Bernoulli with $\sigma = -1$ and $\tau = 2$

pp. 2526–2527). Since there is no basin of attraction for an isle of Eden, the reader has to test all 2^{27} distinct initial bit strings, with $L = 27$, to verify that the above bit string is indeed an isle of Eden.

References

S. Albeverio, V. Jentsch, H. Kantz (eds.), *Extreme Events in Nature and Society* (Springer, Berlin, 2006)

K.T. Alligood, T.D. Sauer, J.A. Yorke, *Chaos: An Introduction to Dynamical Systems* (Springer, New York, 1996)

L.O. Chua, V.I. Sbitnev, S. Yoon, A nonlinear dynamics perspective of Wolfram's new kind of science. Part IV: From Bernoulli shift to $1/f$ spectrum. Int. J. Bifurcation Chaos **15**(4), 1045–1183 (2005a)

L.O. Chua, V.I. Sbitnev, S. Yoon, A nonlinear dynamics perspective of Wolfram's new kind of science. Part V: Fractals everywhere. Int. J. Bifurcation Chaos **15**(12), 3701–3849 (2005b)

L.O. Chua, O.L. Pazienza, V.I. Sbitnev, J. Shin, A nonlinear dynamics perspective of Wolfram's new kind of science. Part IX: Quasi-ergodicity. Int. J. Bifurcation Chaos **18**(9), 2487–2642 (2008)

R.L. Devaney, *A First Course in Chaotic Dynamic Systems: Theory and Experiment* (Addison-Wesley, Reading, 1992)

B.M. Garay, L.O. Chua, Isles of Eden and the ZUK theorem in $\mathbb{R}$. Int. J. Bifurcation Chaos **18**(10), 2951–2963 (2008)

M.W. Hirsch, S. Smale, *Differential Equations, Dynamical Systems, and Linear Algebra* (Academic Press, New York, 1974)

D. Kaplan, L. Glass, *Understanding Nonlinear Dynamics* (Springer, New York, 1995)

K. Mainzer, Real numbers, in *Numbers*, ed. by H.D. Ebbinghaus, H. Hermes, F. Hirzebruch, M. Koecher, K. Mainzer, J. Neukirch, A. Prestel, R. Remmert (Springer, New York, 1990) (German editions: 1983 1st edition, 3rd edition 1992), chapter 2

K. Mainzer, *Thinking in Complexity. The Computational Dynamics of Matter, Mind, and Mankind*, 5th edn. (Springer, Berlin, 2007a)

K. Mainzer, *Der kreative Zufall. Wie das Neue in die Welt kommt* (C.H. Beck, München, 2007b)

K. Mainzer (ed.), Complexity. Eur. Rev. (Academia Europaea) **17**(2), 219–452 (2009)

E.F. Moore, Machine models of self-reproduction. Proc. Symp. Appl. Math. **14**, 17–33 (1962)

H. Nagashima, Y. Baba, *Introduction to Chaos* (Institute of Physics Publishing, Bristol, 1999)

H. Poincaré, *Les Méthodes Nouvelles de la Méchanique Céleste I–II* (Gauther-Villars, Paris, 1897)

M. Schröder, *Fractals,Chaos, Power Laws* (W.H. Freeman & Co, New York, 1991)

L. Shilnikov, A. Shilnikov, D. Turaev, L. Chua, *Methods of Qualitative Theory in Nonlinear Dynamics I–II* (World Scientific, Singapore, 1998)

T. Toffoli, N. Margolus, Invertible cellular automata: a review. Phys. D **45**, 229–253 (1990)

Chapter 6
Time in the Universe of Cellular Automata

An examination of Figs. 10, 11, 12, 13 and 14 at the end of the last chapter shows that, except for period-k isle of Eden bit strings (Figs. 12, 14b), all attractors of the cellular automaton 62 have a non-empty basin of attraction with several gardens of Eden. Therefore, given any bit string on an attractor, it is impossible to retrace its dynamics in *backward time* to find where it had originated in the transient regime. Unlike in ordinary differential equations used in modeling dynamical systems, it is impossible, for most rules of cellular automata, to retrace its *past history on the attractor*. This observation leads us to exciting and deep insights in the concept of time with respect to the universe of cellular automata and physics (Chua et al. 2006; Mainzer 2002; Sachs 1987).

Time Reversal Test of Cellular Automata

How can time reversibility of cellular automata be represented and tested? (Kari 1996; Morita 1989; Toffoli 1977). In Fig. 1, the top pattern shows the evolution under rule 62 from an initial bit string of 63 bits 110110110 … 110110110 on a period-3 isle of Eden, corresponding to the nine-bit isle of Eden shown in Fig. 12a in Chap. 5. There is no transient in the evolution dynamics because the first three rows of Fig. 1 are repeated periodically. Obviously, this is a *period-3 orbit*. In addition, these three rows constitute an isle of Eden, because the subsequence 110 comes from the nine-bit period-3 isle of Eden in Fig. 12a. Now, we consider the *bilateral twin* $118 = \mathbf{T}^{\dagger}(62)$ of 62, where $\mathbf{T}^{\dagger}$ is the *left–right transformation*. The left–right transformation $\mathbf{T}^{\dagger}$ is a member of the four-element *Klein Vierergruppe*, the "holy grail" of symmetry in the universe of cellular automata. A fundamental property of $\mathbf{T}^{\dagger}$ is that any rule N and its bilateral twin $\mathbf{T}^{\dagger}(N)$ are *globally equivalent* to each other, and hence have identical dynamic behavior.

K. Mainzer and L. Chua, *The Universe as Automaton*, SpringerBriefs
in Complexity, DOI: 10.1007/978-3-642-23477-4_6, © The Author(s) 2012

Fig. 1 An invariant set via
rule 62 satisfies the time
reversal test

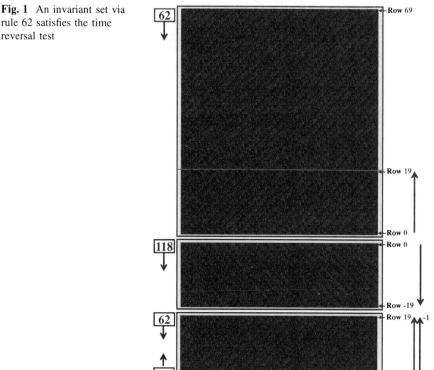

The last row of the top pattern in Fig. 1 (identified as row 0) is chosen as initial
state of the bilateral twin 118 of cellular automaton 62. Iterating this bit string 19
times, we obtain the second pattern in Fig. 1. This 20-row pattern, identified on the
right margin from row 0, -1, -2, ..., -19, is a *mirror image* of the last 20 rows of
the upper pattern, identified by the corresponding row numbers 0, 1, 2, ..., 19.

We can rigorously verify that these two 20-row patterns are exact mirror images
in the following way: The last 20 rows of the upper pattern are reproduced in the
bottom of Fig. 1. Then, the 20-row pattern of the middle pattern is rotated about
row 0 by 180°. In the next step, the resulting flip pattern is superimposed at the
bottom in such a way that the two rows 0 of each pattern are aligned. An examination of the resulting *time reversal comparison pattern* in Fig. 1 shows no difference from the top pattern's original 20-row pattern. Our choice of 20 was
completely arbitrary. In general, it is possible to apply the bilateral twin rule to any
point on an orbit, and successfully recover its past. This procedure is called *time
reversal test.*

Another *period-3 isle of Eden* of cellular automaton 62 is shown in Fig. 2. We
consider the upper pattern showing the evolution under rule 62 from the initial 63
bit string 0001011110000101111 ... 000101111 on a period-3 isle of Eden

Fig. 2 An invariant set via rule 62 violates the time reversal test

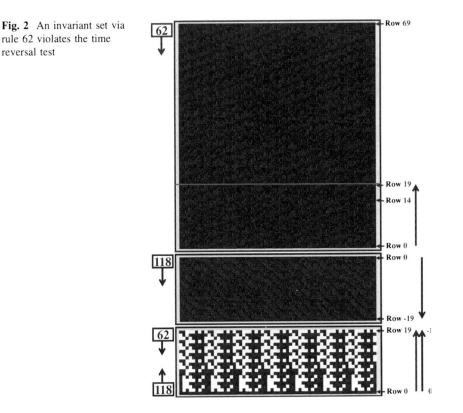

corresponding to the nine-bit isle of Eden in Fig. 12b. Just as in the last example, rule 62 spawns a period-3 invariant set from this initial bit string, without any transients. In a *time reversal test*, the last 20 rows of the top pattern in Fig. 2 are chosen for setting up the time reversal comparison pattern. Applying rule 118 to the last row 0 of the top pattern, we obtain the second pattern in Fig. 2. Copying the last 20 rows of the top pattern to the bottom of Fig. 2 and superimposing the flip pattern of the second 20-row pattern generated by rule 118, we find the results do not coincide with each other. The deviations are indicated by white pixels. It follows that this particular period-3 isle of Eden is *not time-reversible* in the sense that the past evolution of the invariant set under rule 62 cannot be retrieved via its *bilateral twin rule* 118.

Time Reversibility and Arrow of Time

Time reversal tests can be applied to all kinds of *attractors* and *invariant orbits*. Rule 62 has time-irreversible period-3 attractors that violate the time reversal test and time-reversible Bernoulli shift attractors. The last example (Fig. 1) demonstrates that a time-reversible attractor need not be periodic, assuming $I \rightarrow \infty$. There are 170 rules which are endowed with *time-reversible attractors*. Since the remaining 86 rules do not have robust time-reversible attractors, it is impossible to retrieve the past of any bit string lying on their attractors. Inspired by the notion of the direction of time from irreversible thermodynamics, each *time-irreversible attractor* may be understood as an *arrow of time* of the dynamic evolution on the attractor (Chua 2006; Mainzer 2002; Zeh 2007).

In general, an *attractor* $\Lambda(N)$ or an *invariant orbit* $\Lambda(N)$ of cellular automaton N is said to be *time-reversible* iff any k consecutive bit strings $\overrightarrow{x_1}, \overrightarrow{x_2}, \ldots, \overrightarrow{x_k}$ belonging to $\Lambda(N)$ can be completely retrieved by applying the *bilateral twin automaton* $N^{\dagger} = \mathbf{T}^{\dagger}(N)$ of automaton N to the last bit string $\overrightarrow{x_k}$ for a total of k iterations. It follows that the equations

$$\overrightarrow{x_{k-1}} = N^{\dagger}\left(\overrightarrow{x_k}\right)$$

$$\overrightarrow{x_{k-2}} = N^{\dagger}\left(N^{\dagger}\left(\overrightarrow{x_k}\right)\right)$$

$$\overrightarrow{x_{k-3}} = N^{\dagger}\left(N^{\dagger}\left(N^{\dagger}\left(\overrightarrow{x_k}\right)\right)\right)$$

$$\vdots$$

$$\overrightarrow{x_1} = N^{\dagger}\underbrace{\left(N^{\dagger}\left(N^{\dagger}\ldots\left(\overrightarrow{x_k}\right)\right)\right)\ldots\right)}_{k-2 \text{ times}}$$

are satisfied if, and only if, the *time reversal test* is satisfied. An *attractor* $\Lambda(N)$ or an *invariant orbit* $\Lambda(N)$ is said to be *time-irreversible* iff $\Lambda(N)$ is not time-reversible. A *cellular automaton* with rule N is said to be *reversible* iff all attractors $\Lambda(N)$ or all invariant orbits $\Lambda(N)$ of N are time-reversible. A *cellular automaton* with rule N is said to be *time-irreversible* iff all robust attractors $\Lambda(N)$ of N are time-irreversible. An *attractor* is called *robust* iff it can be observed by applying some random initial bit strings.

Time Reversibility and Invertibility

In Chap. 5 Fig. 5, each attractor was characterized by a *forward time-1 return map* $\rho_1[N]$ on the left, and a *backward time-1 return map* $\rho_1^{\dagger}[N]$ on the right. By definition, each return map must have an *inverse*. If the forward return map $\rho_1[N]$ and the backward return map $\rho_1^{\dagger}[N]$ are symmetrical with respect to the main diagonal in Fig. 5 with $\rho_{\tau}[N] = \left[\rho_{\tau}^{\dagger}[N]\right]^{-1}$, then these time-1 return maps were said to be *invertible*. Geometrically (e.g., in Fig. 5), the attractor representation on the left, and its corresponding attractor representation on the right are mirror images of each other with respect to a mirror placed along the main diagonal of each return map. There are 146 rules of cellular automata with at least one invertible attractor.

In general, it can be proved that an attractor $\Lambda(N)$ or an invariant orbit $\Lambda(N)$ of cellular automata N is *time-reversible* if its associated forward (or backward) time-1 return map is *invertible*. In short, invertibility implies time-reversibility, but not vice versa. Thus, all 146 invertible cellular automata have time-reversible attractors. In particular, all 80 invertible Bernoulli shift rules are time-reversible. The other 28 Bernoulli shift rules (among 108) are time-irreversible. But, there are 24 non-invertible and non-bilateral period-1 rules which are nevertheless time-reversible (Chua et al. 2006).

An important contribution of our analytical approach to cellular automata is that the concepts of *time-reversibility* and *invertibility* are rigorously clarified in terms of attractors of local rules. The analytical approach is beyond Wolfram's remarks on the time arrow (Wolfram 2002). The unsuccessful attempts to view time-reversible cellular automata as an extension of time reversal theory from physics, fail, in view of the ubiquitous presence of multiple attractors in cellular automata. Because the same local rule, such as $N = 62$, is endowed with a *time-reversible period-3 isle of Eden* and many *time-irreversible period-3 isles of Eden*, as well as numerous *period-3 attractors* which are *not time-reversible*, any *theory of time-reversibility* for cellular automata must be expressed in terms of *attractors*. Each attractor of a local rule can be classified as either time-reversible or time-irreversible. The *time-reversal test* permits carrying out such tests efficiently on a computer. The test can also be used to identify the end of the transient regime from any initial bit string, for any local rule.

Philosophically speaking, cellular automata generate worlds with different regions with *arrows* or *symmetry of time*. Each rule endowed with at least one time-reversible attractor, or invariant orbit, is counted as a time-reversible rule, although they are not strictly-time-reversible. Out of 256 types of cellular automata, there are 86 time-irreversible rules and 170 rules, which harbor *time-reversible attractors* (symmetry of time). They include all 69 *period-1 rules* (29 are bilateral and invertible, 16 are nonbilateral and invertible and 24 are nonbilateral and noninvertible), 17 *period-2 rules* (all are invertible and bilateral), and 84 Bernoulli σ_τ-*shift rules* (all are invertible and nonbilateral) (Chua et al. 2006). A fundamental characteristic of a *time-reversible attractor* $\Lambda(N)$ of local rule N is that the *past* of any orbit on $\Lambda(N)$ can be uniquely retrieved by iterating its *bilateral twin rule* $N^\dagger = \mathbf{T}^\dagger(N)$ in forward time. Conversely, the past of $N^\dagger$ can be uniquely retrieved by iterating N in forward time. In this sense, the notion of *past* and *future time* is entirely relative. For each time-reversible attractor $\Lambda(N)$ of a local rule N, its associated bilateral twin attractor $\Lambda^\dagger(N) = \mathbf{T}^\dagger(\Lambda(N))$ can be considered a kind of *time machine*.

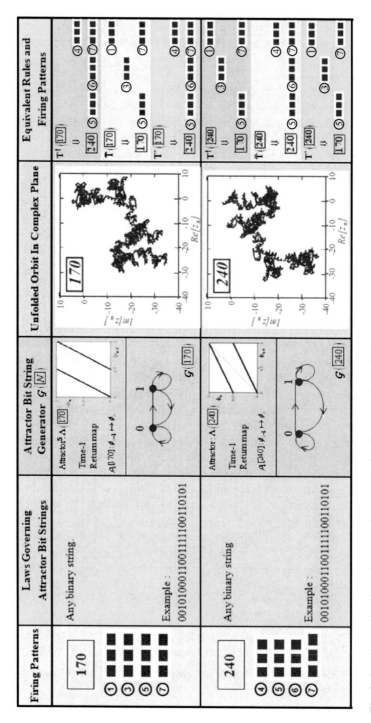

Fig. 3 Random walks of the time-reversible Bernoulli-shift rules 170 and 240

Random Walks, Time and Cellular Automata

To end this chapter, let us come back to the amazing connection between *randomness* and *cellular automata* and the concept of time. In Chap. 5, we have already discussed the relationship of Bernoulli rule 240 and its inverse rule 170 with randomness in *coin toss experiments*. In orbit unfolding plots, the nonlinear dynamics of a local rule N can be illustrated by a walk in the *complex plane*. In this case, each plot is calculated from the *unfolding formula* $z_{n+1} = z_n + \exp(2\pi i\phi_n)$ where z_n is a complex number, $\phi_n = \sum_{k=0}^{I} 2^{-(k+1)}x_k^n$, and x_k^n is the kth component of the bit string $\vec{x}^n = \{x_0^n\, x_1^n\, x_2^n \ldots x_I^n\}$ at the nth iteration under rule N. The *orbit unfolding plots* in Fig. 3 are calculated with $I = 1359$ bit string and iterating over 1366 generations in the complex plane with $\text{Im}(z_n)$ and $\text{Re}(z_n)$ as coordinates. For illustration, a *directed graph* $\mathcal{G}(N)$ is also given along with the bit-string laws for each $\Lambda(N)$, which is no more than a compact algorithm for generating bit strings belonging to $\Lambda(N)$. Directed graphs can represent *complex networks* in different fields of application from electronic circuits to cellular networks (Boccaletti et al. 2006). Thus, they connect models of cellular automata with complex networks.

The two unfolded plots of the *time-reversible* and *invertible Bernoulli shift rules* 170 and 240 in Fig. 3 are reminiscent of *random walks* in *probability theory*. Actually, the unfolding formula is nothing more than a version of an algorithm for illustrating an *ideal coin toss* as a *random walk* (Feller 1950). This interpretation is consistent with the result (mentioned in Chap. 5) that the *Bernoulli map* $\phi_{n+1} = 2\phi_n$ mod 1 is a model of an ideal coin toss. There are two other rules that also exhibit a random walk orbit unfolding plot, namely, rules 15 and 85. These two rules are *globally equivalent* and *time-reversible*.

Each cellular automaton can be characterized by a variety of attractors as forming different possible worlds with their own specific laws. Nevertheless, once the laws governing the attractor bit strings of a rule N are given, it is trivial to derive the laws governing the three other globally equivalent rules of left–right transformation $\mathbf{T}^\dagger(N)$, left–right complementation $\mathbf{T}^\star(N)$, and global complementation $\overline{T}(N)$ from Chap. 4. The laws governing the attractor bit strings and the orbit unfolding plot for rule N and its bilateral twin rule $N^\dagger = \mathbf{T}^\dagger(N)$ are identical. To obtain the corresponding results for rules $\overline{N} = \overline{T}(N)$ and $N^\star = \mathbf{T}^\star(N)$, one simply must only change 0 to 1, and vice versa, and rotate the orbit unfolding plot by $180°$ about the abscissa $\text{Im}(z_n) = 0$.

References

S. Boccaletti, V. Latora, Y. Moreno, M. Chavez, D.-U. Hwang, Complex networks: structure and dynamics. Phys. Rep. **424**(4–5), 175–308 (2006)

L.O. Chua, V.I. Sbitnev, S. Yoon, A nonlinear dynamics perspective of Wolfram's new kind of science Part VI: from time-reversible attractors to the arrow of time. Int. J. Bifurcat. Chaos (IJBC) **16**(5), 1097–1373 (2006)

W. Feller, *An Introduction to Probability Theory and Its Applications I* (Wiley, New York, 1950)

J. Kari, Representation of reversible cellular automata with block permutation. Math. Syst. Theory **29**(1), 47–61 (1996)

K. Mainzer, *The Little Book of Time* (Copernicus Books, New York, 2002)

K. Morita, M. Harao, Computation universality of one-dimensional reversible (injective) cellular automata. Trans. IEICE **E 72**, 758–762 (1989)

R.G. Sachs, *The Physics of Time Reversal* (University of Chicago, Chicago, 1987)

T. Toffoli, Computation and construction universality of reversible cellular automata. J. Comput. Syst. Sci. **15**, 213–231 (1977)

S. Wolfram, *A New Kind of Science* (Wolfram Media, Champaign Il, 2002)

H.-D. Zeh, *The Physical Basis of the Direction of Time*, 5th edn. (Berlin, Springer, 2007)

Chapter 7
Matter in the Universe of Cellular Automata

In the universe of cellular automata, one can identify many concrete concepts and examples that mimic concepts and phenomena of matter in the classical, quantum, and relativistic world of physics. Historically, quantum theory started with *Bohr's atomic model* of an atomic nucleus and discrete orbits of electrons, which remind us of the planetary models of antiquity. In the world of cellular automata, the discrete electron orbits around the nucleus are realized by isles of Eden (Fig. 12, Fig. 14b in Chap. 5). But, Bohr's model was only a rough approximation to the real quantum world. Because of its simplicity and central symmetry, it is still used as an illustration. Bohr's symmetry is only an approximate model. But, the exact *symmetries of the quantum world* lie deeper in the mathematical structure of transformation groups.

Symmetries in the Universe of Physics

In geometry, figures or bodies are called symmetrical when they possess common measures or proportions. Thus the Platonic bodies can be rotated and turned at will without changing their regularity. Plato himself was deeply convinced that the universe was constructed by the five regular bodies of Euclidean geometry. Today, models of Platonic bodies belong to the toy world of basic geometry in schools. The *Boolean cube* is the *Platonic body* in the universe of cellular automata. Mirror and rotational symmetries in the toy world of cellular automata were derived from the Boolean cube.

In general, similarity transformations leave the geometric form of a figure unchanged, for example the proportional relationships of a circle, equilateral triangle, rectangle, etc. are retained, although the absolute dimensions of these figures can be enlarged or decreased. Therefore one can say that the form of a figure is determined by the similarity transformations that leave it unchanged (*invariant*).

K. Mainzer and L. Chua, *The Universe as Automaton*, SpringerBriefs
in Complexity, DOI: 10.1007/978-3-642-23477-4_7, © The Author(s) 2012

In mathematics, a similarity transformation is an example of an automorphism. In general, an *automorphism* is the mapping of a set (such as points, numbers, or functions) onto itself that leaves unchanged the structure of this set (for example the proportional relations in Euclidean space). Automorphisms can also be characterized algebraically in this way: (1) Identity I that maps every element of a set onto itself, is an automorphism. (2) For every automorphism T an inverse automorphism T^{-1} can be given, with $T \cdot T^{-1} = T^{-1} \cdot T = I$. (3) If S and T are automorphisms, then so is the successive application $S \cdot T$. A set of elements with a composition that fulfills these three axioms is called a group. The symmetry of a mathematical structure is determined by the group of those automorphisms that leave it unchanged (invariant) (Mainzer 1996).

Symmetry transformations can be classified in two classes: continuous and discrete transformations. By definition, a symmetry transformation is said to be *continuous* if the set of parameters, which are necessary to describe the transformation, range over a continuous set of values. Examples of continuous transformations are translation in space, rotation around a given axis, and translation in time. These symmetry transformations are global, because once the transformation of a given point in space has been fixed, then the transformation at all other points in space is also fixed. Basic principles of physics like the conservation of linear or angular momentum, and conservation of energy result from the symmetry properties of the interactions under global continuous space and time transformations. According to *Emmy Noether's theorem*, a Lagrangian theory possesses N conserved quantities, if the theory (i.e., the Lagrangian function) is invariant under an N-parameter continuous transformation. Noether's theorem is not only a cornerstone of classical physics, but of quantum physics as well.

A *discrete symmetry transformation* is described by parameters ranging over a discrete set of values. Examples are symmetry operations that leave a crystal unaffected by reflections through planes, inversions with respect to a centre point, and rotations around a given axis with angles $2\pi/n$ (where $n = 2, 3, 4$ or 6) corresponding to the periodicity of the crystal lattice. There are three discrete transformations in physical systems—the charge conjugation C, the parity transformation P, and the time reversal T:

In a *charge conjugation operation* C: $q_\alpha \mapsto -q_\alpha$ all charges q_α change sign. For example, in elementary particle physics, all the particles of a system are replaced by their antiparticles.

The *parity transformation* P: $\mathbf{r} \mapsto -\mathbf{r}$ corresponds to a space inversion relative to a point. In a system of Cartesian coordinates, a point with coordinates (x, y, z) transforms into $(-x, -y, -z)$ under the parity operation. The position vector $\mathbf{r}$ changes sign under a space inversion.

The *time reversal operation* T: $t \mapsto -t$ corresponds to the inversion of the time variable t. The laws of physics are invariant with respect to T. Symmetry of time means that it is physically impossible to distinguish between forward and backward motion in time.

In *classical physics*, physical systems are invariant with respect to each individual transformation of this type, but, in general, that is not true in quantum field theory. Thus, in classical physics, it is trivial that systems are also invariant with respect to the combination of parity, charge, and time-reversal transformations, but, in general, not in quantum field theory. In the famous *CPT-theorem*, quantum theory of fields requires the *invariance* of the fields and interactions under the combined transformations of the three operations *CPT*. The *CPT*-theorem was proved by Wolfgang Pauli in 1957. If one of the three symmetries is violated, then, according to the *CPT*-theorem, one of the other two symmetries must also be violated. For example, the violation of parity *P* requires that *C* or *T* be violated. If the invariance under the combination of two transformations holds, then the invariance under the third transformation must also hold. For example, the invariance under *CP* implies the invariance under *T* and vice versa. The decay of the elementary particles named kaons is the only known example of *time violation T* which is enforced by a CP-violation. In addition, the *CPT* invariance implies that the masses and the lifetime of a particle are identical to those of its antiparticle. *CPT* invariance has been empirically confirmed to a very high precision.

Before 1956, it was assumed that *parity* was a fundamental symmetry of physical processes. In 1956, Tsung Dao Lee and Chin Ning Yang examined the question of whether processes driven by the *weak interaction* would distinguish between left and right. Their famous experiments performed with the beta decay of ^{60}Co, and the weak decays of pions and muons, $\pi^+ \mapsto \mu^+ + \nu_\mu$ and $\mu^+ \mapsto e^+ + \nu_e + \bar{\nu}_\mu$ not only provided empirical support to the suggestions of Lee and Yang but also showed that parity violation was an universal property of the weak interaction (Doncel et al. 1987).

Symmetry concepts play a central role in physics. The invariance and covariance properties of a system under specific symmetry transformations can either be related to the *conservation laws* of physics or be capable of establishing the structure of the *fundamental interactions*. This is the most essential aspect of symmetry, because it concerns the basic principles of physics and the interactions themselves, and not just the properties of a particular system. We will come back to these aspects of physical symmetry with respect to automata in the last chapter.

Symmetries in the Universe of Automata

Understanding symmetries in physics needs a deep knowledge of physical fundamentals. How can the fundamental role of symmetries be illustrated in the toy world of cellular automata? Obviously, there are analogies between discrete symmetries and well-known transformations in the universe of cellular automata: The *left–right transform* $\mathbf{T}^\dagger$ in cellular automata is analogous to the parity operator *P* in physics, and the *global complementation* $\overline{\mathbf{T}}$ in cellular automata is analogous to the charge conjugation operator *C* in physics. Then, the *left–right complementation* $\mathbf{T}^\star$ in cellular automata is analogous to the simultaneous left–right and positive charge-negative charge operator (*CP* mirror) in physics. In *quantum field theory*, the CP-mirror

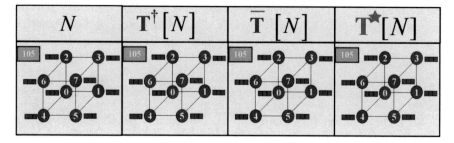

Fig. 1 Symmetric rules analogous to photons (rule 105)

corresponds to the simultaneous left–right and particle-antiparticle operation of elementary particles. In a rigorous sense, analogies with the quantum world need a quantum version of cellular automata. We will discuss this approach in the last section of this chapter. Therefore, the following analogies are only hypothetical. But at least, at this stage, one gets a first glance of symmetries using simple tools in the toy world of cellular automata, and which can be understood even by laymen. In this sense, the forward and backward transformations $T_n(N)$ and $T_{-n}(N)$ under rule N in cellular automata are analogous to the time-reversal operation T in physics.

If you envisage the 256 rules N of cellular automata as matter, then their global complements $\overline{N} = \overline{T}(N)$ can be considered as *antimatter*. An example is 110 as the antimatter of 137. There are 16 rules which are self-complements of each other with $N = \overline{T}(N)$. They can be considered the analog of a photon, which is its own antiparticle. With $N = T^\dagger(N)$, $N = \overline{T}(N)$, and $N = T^\star(N)$ for $N = 105$, we can even find an analog of the symmetry rules of the photon (Fig. 1), in the sense that these particles do not change under particle-antiparticle transformation C, left–right transformation P, nor with the CP-mirror.

The *pair annihilation* and *production* of matter and antimatter in particle physics can be mimicked by appropriate choice of initial states from time-reversible rules (Chua et al. 2006). For example, rule 184 shows the collision of a double-stream of red pixels (mimicking an electron track) on the right, with a symmetrical double-stream of blue pixels (mimicking a positron track) on the left, thereby annihilating each other, resulting in a checkerboard pattern (mimicking the emission of gamma radiation) (Fig. 2a). By applying an excitation (simulated by pixels enclosed with the green rectangle) to the otherwise checkerboard pattern above it (mimicking the physical vacuum), by the associated bilateral twin rule $226 = T^\dagger(184)$, we find the spontaneous generation of a double-stream of red pixels on the right (mimicking an electron) and a symmetrical double stream of blue pixels (mimicking a positron) on the left. The annihilation and pair-production phenomenon in physics is decreed by quantum mechanics, but in cellular automata, we can easily explain how they occur by examining the truth table of local rule 184.

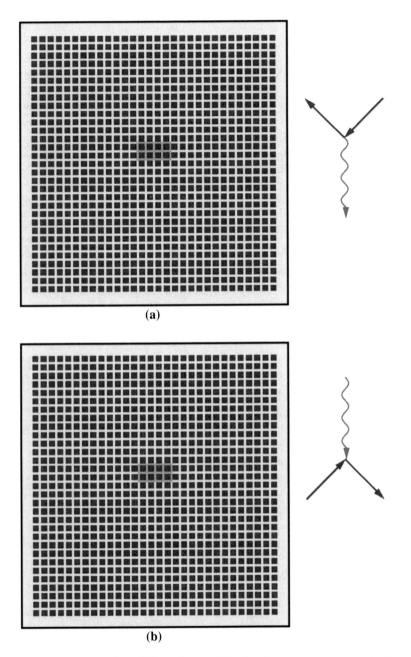

Fig. 2 Matter-antimatter pair annihilation by rule 184 (a) and matter–antimatter production by rule 226 (b) with Feynman diagrams

There is some similarity of the electron–positron annihilation and generation process depicted by the corresponding *Feynman diagrams* (Feynman 1949). It is amazing that only two adjacent red (or blue) pixels are needed to emulate an electron or positron, respectively, to distinguish it from the checkerboard background (emulating the vacuum condensate $\langle 0|e^+ \ e^-|0\rangle$ from quantum physics). If we identify rules 184 and 226 as the same rule, via their global equivalence, then one can even think of the positron as an electron traveling backwards in time. We can also interpret the red stream and the blue stream as electron–electron scattering. But, of course, these are only illustrations of the Feynman diagrams by (classical) cellular automata. To model quantum processes, one needs quantum cellular automata.

There are two additional Bernoulli rules that exhibit the same space–time diagram as that of rule 184, namely rules 56 and 57. However, there is a subtle difference between the $\mathbf{T}^\star = \overline{\mathbf{T}} \cdot \mathbf{T}^\dagger$ transformation (an element of the Vierergruppe $\mathcal{V}$) of rule 56 and that of rules 184 and 57; namely, $\mathbf{T}^\star(56) = 185 \neq 56$, but $\mathbf{T}^\star(184) = 184$ and $\mathbf{T}^\star (57) = 57$. In other words, whereas both rules 184 and 57 possess the $\mathbf{T}^\star = \overline{\mathbf{T}} \cdot \mathbf{T}^\dagger = \mathbf{T}^\dagger \cdot \overline{\mathbf{T}}$ symmetry, in contrast, rule 57 violates the $\overline{\mathbf{T}}$ symmetry. It follows that even though $\mathbf{T}^\dagger(184) \neq 184, \mathbf{T}^\dagger(57) \neq 57, \overline{\mathbf{T}}(184) \neq 184$, and $\overline{\mathbf{T}}(57) \neq 57$, the pair-annihilation and pair-production process of both rules 184 and 57 mimic the *CP-symmetry property* from quantum electrodynamics, where the parity operator P is analogous to symmetry transformation $\mathbf{T}^\dagger$ of Klein's Vierergruppe $\mathcal{V}$, and the *particle-antiparticle conjugation operator C* is analogous to the $\overline{\mathbf{T}}$ symmetry. In contrast, rule 56 mimics the *CP-violation* phenomenon observed in the kaon elementary particle.

Any physical law that remains unchanged after simultaneous inversion of charge C, parity P, and time T, is said to exhibit *CPT symmetry*. In the world of elementary particles, the physics of particles described in a right-handed coordinate system is the same as the physics of antiparticles described in a left-handed coordinate system. Rule 56 of cellular automata is the physical analog of *CPT* symmetry. Rule 56 violates *CP* symmetry, but exhibits *CPT* symmetry. We know $\mathbf{T}^\star(56) = 185$ is not equal to 56. However, it is time-reversible for all bit strings belonging to an attractor of 56. Hence, 56 exhibits *CPT* symmetry on its attractors (Chua et al. 2006). As with *CP* violation, T violation occurs in certain weak interactions, such as kaon decay.

Notice that the analogies of CPT-symmetry and CP violation are only metaphors or hypothetical analogies with cellular automata and are not models of physics in a rigorous sense. They do *not* explain the *spontaneous process of symmetry breaking* leading to the emergence of new particles and quantum forces in *quantum field theory* (Mainzer 2005a, b). Nevertheless they illustrate that the physical phenomena of symmetries and symmetry breaking are not strange, but understandable even in the toy world of automata.

Expansion in the Universe of Physics and Automata

At least as a metaphor, we can also imagine a contracting and expanding toy universe of automata. Here, the singular point where the expansion begins is analogous to the *Big Bang* event in cosmology. Rule 30 can mimic the Big Bang. In addition, one can associate "isles of Eden" from cellular automata with the frozen time, which is a well-known effect of black holes in cosmology. But the analogies with physical time are actually more sophisticated. Time-reversible attractors and time-reversible isles of Eden in cellular automata cannot simply be viewed as an extension of time-reversibility in physics. The reason is the *ubiquitous presence of multiple attractors* and *isles of Eden* in cellular automata. Because a local rule, for example 62, is endowed with time-reversible period-3 isle of Eden and many time-irreversible period-3 isles of Eden, as well as numerous period-3 attractors that are not time-reversible, any meaningful theory of time-reversibility for cellular automata must be couched in terms of attractors.

A fundamental characteristic of a time-reversible attractor $\Lambda(N)$ of rule N is that the past of any orbit on $\Lambda(N)$ can be uniquely retrieved by iterating its bilateral twin rule $N^\dagger = \mathbf{T}^\dagger(N)$, in forward time. Conversely, the past of $N^\dagger$ can be uniquely retrieved by iterating N in forward time. In this sense, any attractor has its own time, reminding us of Einstein's relativity of time. It is even more intriguing to observe that for each time-reversible attractor $\Lambda(N)$ of a rule N, its associated bilateral twin attractor $\Lambda^\dagger(N) = \mathbf{T}^\dagger(\Lambda(N))$ is in fact a kind of time machine. This concept seems to be science fiction, but does not violate the laws of physics. Again, the analogies with physics are only hypothetical metaphors. But they underline that these concepts are not strange, but understandable in the toy world of automata.

A remarkable phenomenon exhibited by the bilateral twin rules 184 and 226 = $\mathbf{T}^\dagger(184)$ is that their respective transient regimes are *time-reversible*, a property which we have so far restricted to evolution on attractors. Indeed, the upper space–time diagram in Fig. 2a ending at the middle horizontal line segment of the small green rectangle, represents a transient regime of 184. Similarly, the lower space–time diagram in Fig. 2b, beginning from the middle horizontal line segment of the small green rectangle, also represents a transient regime of rule 226. The two transient regimes of 184 and 226 can be considered time machines. In this instance, the past of 184 is the future of 226 and vice versa, over the corresponding duration of the transient regimes. It follows that the future and the past cannot be defined absolutely, which again reminds us of Einstein's theory of relativity (Chua et al. 2006; Mainzer 2002).

The metaphors and analogies we have drawn from quantum field theory and theory of relativity illustrate that in the universe of cellular automata, there are examples of computational systems that exhibit similar phenomena, thereby demonstrating that such phenomena are neither counter-intuitive nor strange, but reducible to *symmetry laws* even outside physics. One of the most amazing aspects of these analogies is the fact that very *simple rules* in a toy world of automata lead to high *complexity* which is not even always predictable in the long run. The hard core of this insight is the global equivalence class of automata 110, 124, 137, and 193.

Quantum Matter and Quantum Information

From of physical point of view, the (1-dimensional) cellular automata we introduced in the previous chapters are deterministic dynamical systems when viewed against the background of classical physics. In 1982, Richard Feynman, Nobel prize winner and one of the most influential physicists in the twentieth century, introduced the notion of *quantum computation*. Feynman strongly argued that processes of quantum physics cannot be simulated by computers based on classical physics. Thus, for Feynman, phenomena of the quantum world such as elementary particles could only be modeled by quantum versions of cellular automata (Feynman 1982).

Let us start with a reminder of crucial concepts in the quantum world (Audretsch and Mainzer 1996). One of the fundamental differences from classical physics is the different concept of *states in quantum physics*. Consider a weak light source set up to shine at a pair of detectors. These detectors are sensitive enough for them to emit a signal ('click') when an individual photon arrives. In this experiment, light acts like particles. When the light becomes weaker, fewer, rather than weaker, clicks are observed at the detector.

Assume that a half-silvered mirror is placed in the light beam. Then, quantum physics predicts, and it is confirmed in experiments, that the photons will be detected at one or the other site with equal probability. Classically, this observation is rather strange, for how does the photon decide which way to go? Certain photons must be predisposed to reflect, while others are predisposed to pass through the mirror. In quantum physics, the state of a photon passing the mirror is considered a *superposition* of being simultaneously reflected and not reflected, with both having equal probability. It can exist in the superposition state until it reaches one of the detectors, when it is forced into one partial state or another, i.e., being reflected or not. This expresses one of the most important differences between quantum and classical mechanics: The act of measurement in a quantum system irreversibly changes the system state.

Entanglement is another quantum mechanical phenomenon that cannot be explained by classical physics. In the famous EPR-Paradox (named after Einstein, Podolsky, and Rosen), two separated objects can be correlated in an entangled state. For example, two separated photons may be in an entangled state such that measuring one of them also forces the result of the measurement of the other photon. Classically, this seems to be strange because instantaneous communication about the measurement between the two is needed, contrary to Einstein's relativity postulate that no information can be transmitted faster than the speed of light. In 1964, John Bell proved that correlation of the measurement of both entangled photons is higher than classical statistical physics predicts (Bell 1964). Bell's prediction was experimentally confirmed by A. Aspect et al. in 1982. These results were strong arguments for Feynman's demand that *quantum probabilities* cannot be simulated by *classical probabilistic computers,* which would act like classical nondeterministic Turing machines (Feynman 1982).

In a *quantum computer,* a *superposition* is used as the basic unit of information, called a *quantum bit* (qubit). A bit in a classical computer stores a binary value, either 1 or 0. A qubit is stored as a two-state quantum system such as, for example, an electron that is either spin-up or spin-down, or a photon with either horizontal or vertical polarization. In Dirac's notation, qubits are represented as a ket, where the values 1 and 0 are denoted as $|1\rangle$ or $|0\rangle$. Until it is measured, the qubit is in a superposition of 1 and 0 which is represented by a probability distribution over the values. Although the probability distribution cannot be measured directly, it can take part in computation (Bouwmeester et al. 2000; Mainzer 2007).

Mathematically, a qubit is a unit state vector in a *2-dimensional Hilbert space* with $|1\rangle$ and $|0\rangle$ as orthonormal basis vectors. For each qubit $|x\rangle$, there exist two complex numbers a and b such that $|x\rangle = a|0\rangle + b|1\rangle = \begin{pmatrix} a \\ b \end{pmatrix}$ with $|0\rangle = \begin{pmatrix} 1 \\ 0 \end{pmatrix}$ and $|1\rangle = \begin{pmatrix} 0 \\ 1 \end{pmatrix}$, and $|a|^2 + |b|^2 = 1$. Geometrically, a and b define the angle that the qubit makes with the vertical axis, indicating the probability that the given bit will be measured as 0 or 1.

Similar to a classical register, a register of 3 qubits can store $2^3 = 8$ values. In a quantum computer, these values are in a *superposition,* storing all 8 values at once, with a joint probability distribution across the set of values. Thus, their computation can be realized in one parallel procedure and does not need eight separate procedures. When particular values are read out, the superposition breaks down and is forced into one of the partial states with specific values. In addition, it turns out that physical realizations of qubits are very sensitive to noise and perturbations in the environment. One of the great technical challenges of quantum computers therefore is to shield a superposition with respect to noise and perturbations by the environment.

The Universe of Quantum Cellular Automata

In the previously mentioned 1982 paper on "Simulating Physics with Computer", Richard Feynman already mentioned quantum cellular automata as a possible model for making a universal quantum computer. In 1985, David Deutsch defined the concept of a *universal quantum Turing machine* and extended the *Church-Turing principle* to a quantum Turing machine (Deutsch 1985). But it was John Watrous who was one of the first to come up with a mathematical definition of *quantum cellular automata* in 1995 (Watrous 1995, Horowitz 2008). From an academic point of view, his definition is easy to understand. A classical 1-dimensional cellular automaton is defined by a local deterministic *transition rule* $v_i = N(u_{i-1}, u_i, u_i)$, in accordance with a prescribed *Boolean truth table* of $8 = 2^3$

distinct 3-input patterns (u_{i-1}, u_i, u_i) and corresponding output values v_i after their application. The idea of Watrous' quantum concept is to replace the uniquely determined value v_i with $v_i \in \{0, 1\}$, by a *quantum probabilistic distribution* or quantum amplitude $|x\rangle = a|0\rangle + b|1\rangle$ of states $|0\rangle$ and $|1\rangle$, with complex numbers $a, b \in \mathbb{C}$ and $|a|^2 + |b|^2 = 1$.

The classical update rule N can be illustrated by a map

$$N : \underbrace{\Sigma}_{\text{left}} \times \underbrace{\Sigma}_{\text{old}} \times \underbrace{\Sigma}_{\text{right}} \rightarrow \underbrace{\Sigma}_{\text{new}} (\text{with } \Sigma \text{ as the set of cell states}), \text{where we have}$$

the old states of cells, their left and right neighbors, and the new states.

The quantum update rule N_q assigns quantum amplitudes to every possible transition from old states and their two left and right neighbors to new states by a map

$$N_q : \underbrace{\Sigma}_{\text{left}} \times \underbrace{\Sigma}_{\text{old}} \times \underbrace{\Sigma}_{\text{right}} \times \underbrace{\Sigma}_{\text{new}} \rightarrow \underbrace{\mathbb{C}}_{\text{amplitude}} \text{ with the set } \mathbb{C} \text{ of complex numbers.}$$

Based on these local transition amplitudes, the global transition amplitudes from any given configuration to any other configuration can be computed by $N_Q(u, v) = \Pi_i N_q(u_{i-1}, u_i, u_i, v_i)$. With these amplitudes, a single-time-step-evolution operator T can be introduced in a Hilbert space with configurations as basic elements, i.e., $T|u\rangle = \sum_v N_Q(u, v)|v\rangle$.

In the quantum world, every quantum system must evolve according to some unitary transformation. Without going into mathematical details, a *unitary operator* guarantees that squared amplitudes are preserved, and can be interpreted as probabilities which sum to 1. Quantum cellular automata are called *well-formed* iff they have transition rules N_q giving rise to unitary transformations T. There is a polynomial-time algorithm to decide if a quantum cellular automaton is well-formed or not. But the question of which quantum cellular automata are well-formed is still rather difficult.

Therefore, Watrous suggested a class of quantum cellular automata for which checking well-formedness is generally easier. These so-called partitioned quantum cellular automata are at least as powerful as quantum cellular automata, but are not any more powerful. They can simulate quantum Turing machines and vice versa. *Quantum Turing machines* are analogous to *probabilistic Turing machines,* with the main difference being that the transition of configurations is not assigned to classical probabilities, but to quantum amplitudes. But equivalence has only been proved between quantum Turing machines and the restricted class of partitioned quantum cellular automata. The question of whether non-partitioned quantum cellular automata are more powerful than quantum Turing machines is still open (Van Dam 1996).

The research on quantum cellular automata opens new avenues to modern physics. *Lattice field theory* studies lattice models of quantum field theory with a spacetime that has been discretized onto a lattice (Giles and Thorn 1977). Although most lattice field theories are not exactly solvable, they are interesting in digitized physics because they can be simulated on computers. There are already relations between lattice field theories and string bit models, as well as new concepts of bosonic, fermionic, and supersymmetric quantum cellular automata (McGuigan 2003). We will come back to these aspects of *digitized physics* in the last chapter.

References

J. Audretsch, K.(Hrsg.) Mainzer, *Wieviele Leben hat Schrödingers Katze? Zur Physik und Philosophie der Quantenmechanik*, 2nd edn. (Spektrum Akademischer Verlag, Heidelberg, 1996)

J.S. Bell, On the Einstein-Podolsky-Rosen-Paradoxon. Physics **1**, 195–200 (1964)

D. Bouwmeester, A. Ekert, A. Zeilinger (eds.), *The Physics of Quantum information. Quantum Cryptography, Quantum Teleportation, Quantum Computation* (Springer, Berlin, 2000)

L.O. Chua, V.I. Sbitnev, S. Yoon, A nonlinear dynamics perspective of wolfram's new kind of science Part VI: from time-reversible attractors to the arrow of time. Int. J. Bifurcat. Chaos (IJBC) **16**(5), 1097–1373 (2006)

D. Deutsch, Quantum theory, the Church-Turing principle and the universal quantum computer. Proc. R. Soc. Lond. A **400**, 97–117 (1985)

M.G. Doncel, A. Herrmann, L. Michel, A. Pais, *Symmetries in Physics 1600–1980* (Servei de Publicacions, UAB Barcelona, 1987)

R.P. Feynman, The theory of positrons. Phys. Rev. **76**, 749–759 (1949)

R. Feynman, Simulating physics with computers. Int. J. Theor. Phys. **21**(6–7), 467–488 (1982)

R. Giles, C. Thorn, Lattice approach to string theory. Phys. Rev. D **16**, 366 (1977)

J. Horowitz, An introduction to quantum cellular automata (2008), http://web.mit.edu/joshuah/www/ projects/qca.pdf

T.D. Lee, C.N. Yang, Questions of Parity Conservation in Weak Interactions. Phys. Rev. **104**, 254 (1956)

K. Mainzer, *Symmetries of Nature* (De Gruyter, New York, 1996) (German 1988: Symmetrien der Natur. De Gruyter: Berlin)

K. Mainzer, *The Little Book of Time* (Copernicus Books, New York, 2002)

K. Mainzer, *Symmetry and Complexity: The Spirit and Beauty of Nonlinear Science* (World Scientific, Singapore, 2005a)

K. Mainzer, Symmetry and Complexity in Dynamical Systems. Eur. Rev. Academia Europaea **13**(2), 29–48 (2005b)

K. Mainzer, *Der kreative Zufall. Wie das Neue in die Welt kommt* (C.H. Beck, München, 2007)

M. McGuigan, Quantum cellular automata from lattice field theories (2003), http://arxiv.org/ftp/quant-ph/papers/0307/0307176.pdf

W. Pauli, *Niels Bohr and the Development of Physics* (Pergamon Press, London, 1957)

W. Van Dam, *Quantum Cellular Automata*. Master's thesis. Computing Science Institute. University of Nijmegen, The Netherlands, 1996

J. Watrous, On one-dimensional quantum cellular automata. Proceedings of the 36th Annual Symposium on Foundations of Computer Science. IEEE Computer Society Press, Milwaukee (Wisconsin), 1995

Chapter 8
Life and Brain in the Universe
of Cellular Automata

Historically, in science and philosophy people believed in a sharp difference between "dead" and "living" matter. Aristotle interpreted life as the power of *self-organization* (entelechy) driving the growth of plants and animals to their final form. A living system is able to reproduce itself and to move by itself, while a dead system can only be copied and moved from outside. Life was explained by teleology, i.e., by non-causal ("vital") forces aiming at some goals in nature. In the eighteenth century Kant showed that self-organization of living organisms cannot be explained by a mechanical system of Newtonian physics. In a famous quotation he said that the Newton for explaining a blade of grass was still lacking. Nowadays, children put the same question: How is it possible that complex organisms such as plants, animals, and even humans emerge from the interactions of simple elements such as atoms, molecules, or cells? The concept of cellular automata was the first mathematical model to prove that *self-reproduction* and *self-organization* of complex patterns from simple rules are universal features of dynamical systems. Therefore, the belief in some preprogrammed *intelligent design* is unnecessary.

Self-Organization and Emergence in Cellular Automata

In the nineteenth century the *second law of thermodynamics* described the irreversible movement of closed systems toward a state of maximal entropy or disorder. The law is supported by our everyday experiences. For example, the flow of heat in a closed room tends to average any local differences and ends in an equilibrium state of temperature everywhere in the room. Spontaneous heating at a localized point in a closed room was never observed. But how could one explain the emergence of order in Darwinian evolution of life? Ludwig Boltzmann stressed that living organisms are *open dissipative systems* exchanging matter, energy, and information with their environment, and which do not violate the second law of closed systems. But nevertheless in Boltzmann's statistical interpretation,

K. Mainzer and L. Chua, *The Universe as Automaton*, SpringerBriefs
in Complexity, DOI: 10.1007/978-3-642-23477-4_8, © The Author(s) 2012

the emergence of life could only be a contingent event, a local cosmic fluctuation "at the boundary of the universe".

In the framework of complex dynamical systems, the emergence of life is not contingent, but necessary and lawful in the sense of self-organization (Creutz 1997; Haken and Mikhailov 1993; Mainzer 2007). Only the conditions for the *emergence of life* in the universe (for instance on planet Earth) may be contingent. In general, biology distinguishes ontogenesis (the growth of organisms) from phylogenesis (the evolution of species). In either case we have open complex systems, the development of which can be explained by the evolution of (macroscopic) patterns caused by nonlinear (microscopic) interactions of molecules and cells in phase transitions that are far from thermal equilibrium. It is well known that Turing analyzed a mathematical model of *cellular pattern formation* (Turing 1952). Gerisch and Meinhardt et al. described the growth of an organism (e.g., a slime mould) by dissipative nonlinear equations for the aggregation of cells (Gerisch and Hess 1974). How can these processes be described by cellular automata?

According to the second law of thermodynamics, closed systems tend to average out any randomness inherent in the initial state and must tend to a thermodynamic equilibrium, completely oblivious of the distant initial state. Such systems have an evanescent memory and therefore cannot exhibit any complexity. Viewed in this context, rule 170 as well as all 108 *Bernoulli* σ_τ *shift rules* are truly remarkable, because they are endowed with an infinite memory in the sense that no bit in the initial state that originates from an attractor, or isle of Eden, is ever averaged out in its dynamic evolution, no matter how far it continues into the future. This is true for all 108 Bernoulli rules, following from just one simple rule: Copy your $|\sigma|$ th right neighbor if $\sigma > 0$, or left neighbor, if $\sigma < 0$, and complement it if $\beta < 0$, every τ iterations, to infinity.

With this simple recipe, one can, in principle, easily deliver random binary strings residing on any given attractor of any one of the 108 *Bernoulli* σ_τ-shift rules by following the laws governing the attractor bit strings on the attractor. This insight explains the seeming paradox of how rule 170 can be associated with a coin toss experiment or random walk, as depicted in the orbit unfolding plot, when its Bernoulli map $\phi_{n+1} = 2\phi_n$ mod 1 is deterministic and predictable. This paradox is caused by the fallacious inference asserting that the *Bernoulli coin toss interpretation* implies that rule 170 can be used as a practical random sequence generator. But that is not true. The Bernoulli map merely asserts that if one applies a random binary sequence $\vec{x}(0) = (x_0(0), x_1(0), \ldots, x_{i-1}(0), x_i(0))$ of binary values 0 and 1 to a cellular automaton with rule 170, the machine with this program will spit out the same sequence as its output by shifting each pixel of $\vec{x}(0)$ by one bit to its left. However, although the output of rule 170 is indeed a random binary sequence, it is not generated by rule 170. But rather it is given at the outset as its input.

Computer experiments with cellular automata generate an inexhaustible variety of structures. Many of them remind us of structures emerging in nature by *self-organization*. The *Game of Life* was one of the most widely-known cellular automata which evolves according to extremely simple local rules. Invented by John Conway as an abstraction of real life, it has attracted world-wide interests after Conway and his collaborators proved that not only is the Game of Life capable of universal computation as a Turing machine, but that it is also capable of self-replication, another essential condition for real life to be possible (Gardner 1970, 1971).

In its original version, the Game of Life is played on an infinite grid of cells, like a checker board. Each cell can assume one of two states, alive (coded in red) or dead (coded in blue). The game is initiated by assigning an initial state to each cell on the checker board so that at $t = 0$, the board looks like mosaic of random juxtaposition of red and blue tiles. The game evolves at discrete intervals, that are usually referred to in the Game-of-Life literature as generations, as an abstraction of real-life events where each species in one generation evolves from its parents in the previous generation, or from its parent's parents in their previous generations, etc. From any given initial configuration of states at time t, each cell in the next generation $t + 1$ evolves in accordance with four simple local rules involving only the state of each of its eight nearest neighbors in the plane. The *four local rules* were as follow:

1. *Birth*. A cell that is dead at time t becomes alive at time $t + 1$ only if exactly three of its eight neighbors were alive at time t.
2. *Death by overcrowding*. A cell that is alive at time t and has more than three living neighbors at time t will be dead at time $t + 1$.
3. *Death by exposure*. A cell that is alive at time t and has less than two living neighbors at time t will be dead by time $t + 1$.
4. *Survival*. A cell that was alive at time t will remain alive at time $t + 1$ if and only if it had exactly two or three alive neighbors at time t.

The biological interpretation of these rules may be more or less arbitrary. But, the *principle of self-organizing global patterns* with high *computational complexity* from simple local rules is obviously universal and independent of biochemical ingredients.

Systems Biology and Cellular Automata

The Game of Life was an abstract model to study the emergence of complex structures from simple local rules in a toy world of cellular automata. Modern *systems biology* studies the emergence of complex cellular structures with reference to lab experiments and measurement data (Alon 2006). The mathematical models are *complex networks* and circuits with great similarity to the dynamics of cellular automata (Mainzer 2010). Although systems biology uses mathematical models like those of physics, there seem to be tremendous differences. Structures spontaneously assemble, perform elaborate biochemical functions, and vanish effortlessly when their work is done. How could this be?

Systems biology, like the theory of cellular automata, believes that the biological components at every level of organization interact with each other and thereby may lead to responses or properties that are not explainable by the study of components in isolation. Their primary goal is to capture the emerging properties and to understand how they arise, how they are implemented in the cell or organism, and what consequences ensue if they are altered, be it physiologically, pathologically, or for biotechnical purposes.

We yearn for simplifying principles, but biology is astoundingly complex. Every biochemical interaction is exquisitely crafted, and cells contain networks of many of such interactions. These networks are the result of billions of years of evolution, which works by making random changes and selecting the organisms that survive. Therefore, the structures found by evolution are dependent on historical chance and are laden with biochemical detail that requires special description in many respects.

Despite this complexity, scientists have attempted to find generalizable principles of biology. Actually, general mathematical laws referring to biological networks and circuits are confirmed by lab experiments and measurements. In systems biology, a cell is considered a *complex system* of interacting proteins (Mainzer 2010). Each protein is a kind of *nanometer-size molecular machine* that carries out specific tasks. For example, the bacterium *Escherichia coli* is a cell containing several million proteins of 4,000 different types. In response to changing situations, cells generate appropriate proteins. When, for example, it is damaged, the cell produces repair proteins. The cell thus monitors its environment and calculates the extent to which each type of protein is needed.

This information-processing function, which determines the rate of production of each protein, is realized by *transcription networks*. The environmental states of a cell are represented by special proteins called transcription factors. They are designed to switch rapidly between active and inactive states. Each active transcription factor binds to the DNA to regulate the rate at which specific target genes are read. The genes are "read" (transcribed) into mRNA, which is then translated into proteins, which can act on the environment. Thus, the activities of the transcription factors in a cell can be considered an internal representation of the environment.

Transcription factor proteins are themselves encoded by genes, which are regulated by other transcription factors, which in turn may be regulated by yet more transcription factors (Alon 2003). The set of interactions is called a transcription network (Fig. 1). The transcription network describes all *regulatory transcription interactions* in a cell. In the network, the nodes are genes, and edges represent transcriptional regulation of one gene by the protein product of another gene. A directed edge X → Y means that the product of gene X is a transcription factor protein that binds to the promoter of gene Y to control the rate at which gene Y is transcribed.

The connection between *complex networks* in systems biology and *cellular automata* is now explained, using the example of gene regulation networks. *Gene regulation networks* can be mapped on the sequences of bit strings in cellular automata. The local rule of a cellular automaton defines different *local patterns*

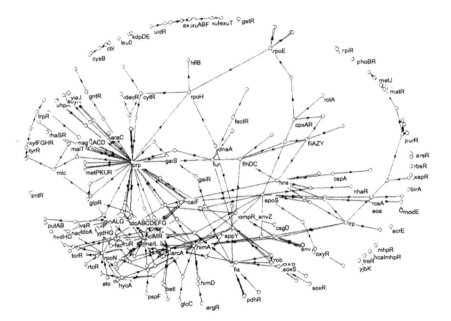

Fig. 1 A transcription network representing about 20% of the transcription interactions in the bacterium *E. coli*

which are followed by one of two binary cellular states (bits) (see Fig. 10 in Chap. 5). These binary bit states are denoted by the colors red and blue. Local patterns followed by a red state are called *firing*, those followed by a blue state are called *quenching*. They generate a sequence of bit strings. If a local pattern of a rule is applied in a sequence of bit strings, it is called *active*, or else *inactive* with respect to that sequence.

In a gene regulatory system, an inactive *firing* local pattern can be identified with an unexpressed gene for an *excitatory* protein molecule, and an inactive "quenching" local pattern can be identified with an unexpressed gene for an *inhibitory* protein molecule. The transition between excitatory and inhibitory protein molecules in a gene regulation network is represented by a sequence of firing and quenching local patterns in a corresponding cellular automaton.

For example, for local rule 62 in Fig. 2b, there are eight possible local patterns of three cells with the five "firing" patterns 1, 2, 3, 4, 5, and the two "quenching" patterns 0, 6, 7. For the period-3 isle of Eden $\Lambda_2(62)$ of local rule 62 in Fig. 2b, the active *firing* patterns consist of 3 and 5, and the active *quenching* patterns consist of just 6. The inactive *firing* patterns consist of 1, 2, and 4, and the inactive *quenching* patterns consist of 0 and 7. Similarly, for the Bernoulli attractor $\Lambda_1(62)$ in Fig. 2a, the active *firing* patterns consist of 1, 3, and 4, and the active *quenching* patterns consist of 6 and 7. The inactive *firing* pattern consists of 2 and the inactive *quenching* pattern consists of 0.

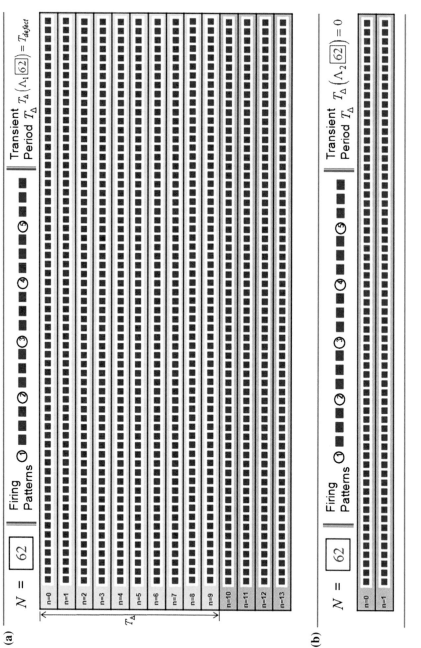

Fig. 2 Dynamic evolution from a generic initial bit for rule 62 on Bernoulli attractor $\Lambda_1(62)$ (**a**) and the period-3 isle of Eden $\Lambda_2(62)$ (**b**)

Just as the dynamical mechanisms leading to a time-2 (right-left) attractor $\Lambda_1(62)$ can be explained and predicted rigorously via the Bernoulli shift law, by invoking only the active *firing* and *quenching* patterns of $\Lambda_1(62)$, so too can attractors associated with gene regulatory networks be explained, and predicted, at least at a conceptual level.

The different tools for cellular automata that are introduced in this booklet can be used to analyze complex networks in systems biology. The transition between excitatory and inhibitory states in gene regulatory networks is illustrated by the pattern formation of bit strings in cellular automata. Directed graphs (c.f. Fig. 3 in Chap. 6) can be used as generators (algorithm) of the bit strings in an attractor. The basins of attraction (c.f. Fig. 11 in Chap. 5) can be represented by basin trees. To get, at least, a glance at the complexity in cellular transcription networks with cellular automata, let us consider a basin tree of the period-3 attractor $\Lambda_3(62)$ (Fig. 3). For $I + 1 = 9$ there are 18 Gardens of Eden (in pink color), each one spawning a sub-basin of the period-3 attractor.

It is a challenge in systems biology to distinguish different patterns in networks. Out of the many possible patterns that could appear in a network, only a few of them are realized in nature. The different structures have particular information-processing functions. The advantages of these functions may explain why the same network structures are repeated by evolution several times in different systems. For example, there are feed forward loops as well as recurring networks. Besides *transcription networks*, we also distinguish *developmental* and *transduction networks*. Sensory transcription networks are designed to rapidly respond to changes in the environment. A developmental transcription network governs the developmental states of cells such as, for example, how an egg develops into a multi-cellular organism. A signal transduction network processes information using interactions between signaling proteins.

Unlike electronic circuits, computations performed by a biological circuit depend on biochemical parameters with concentrations of proteins varying from cell to cell, even if the cells are genetically identical (Kriete et al. 2006). Therefore, systems biologists try to find *biological circuits* with robust designs such that their essential functions are nearly independent of biological parameters. Only robust networks open an avenue to programming cells in synthetic biology for particular medical or technical purposes. These circuits can be mapped on networks representing the dynamics of cellular automata. Systems biology is an exciting application of complex networks that may be modeled by cellular automata (Kayama 2010; Topa 2011).

Brain Research and Cellular Automata

Our cellular automata approach also provides an intriguing bridge for research in neural science on *pattern formation* and *recognition*. There are common concepts in these areas, which can be mapped onto attractors of cellular automata, and vice versa. For example, each attractor $\Lambda(N)$ is hardwired to recognize only a small

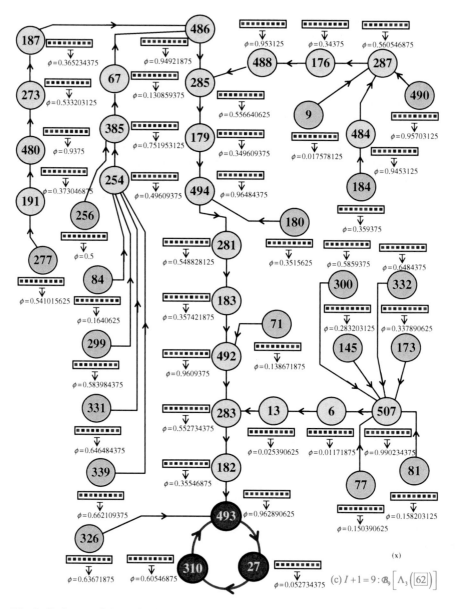

Fig. 3 Basin tree of the period-3 attractor $\Lambda_3(62)$ with nine-bit strings

subset of local *firing* and *quenching* patterns. The excluded patterns are henceforth called inactive local patterns. They too can include *firing* and *quenching* patterns (Fig. 2). From a neural network perspective, an active *firing* local pattern can be identified with an *excitatory* synapse and a *quenching* local pattern can be identified with an "inhibitory" synapse.

A neural network approach which is quite similar to cellular automata (CA) is the concept of *cellular neural networks* (CNN) (Chua 1998). The main idea behind the CNN paradigm is the *local activity principle* which asserts that no complex phenomena can arise in any homogeneous media without local activity. Obviously, local activity is a fundamental property in micro-electronics, where, for example, vacuum tubes and (later) transistors have been locally-active devices in the electronic circuits of radios, televisions, and computers. The demand for local activity in neural networks was motivated by practical reasons of technology (Mainzer 2003). In 1985, John Hopfield theoretically suggested a neural network which, in principle, could realize pattern recognition. But its globally connected architecture was highly impractical to be realized technically in VLSI (very-large-scale-integrated) micro-electronic circuits: The number of wires in a fully connected Hopfield network grows exponentially with the size of the array. A CNN only needs electrical interconnections in a prescribed sphere of influence.

In general, a CNN is a *nonlinear analog circuit* that processes signals in real time. It is a multi-component system of regularly spaced identical ("cloned") units, called cells, which communicate with each other directly only through their nearest neighbors. But the locality of direct connections also allows global information processing to be obtained. Communications between non-directly (remote) connected units are obtained by passing through other units. The idea that complex and global phenomena can emerge from local activities in a network dates back to the paradigm of *cellular automata* (CA). In this sense, the CNN paradigm is a higher development of the CA paradigm taking account of the new conditions in information processing and chip technology. Unlike conventional cellular automata, CNN host processors accept and generate analog signals in continuous time, with real numbers as interaction values. But actually, the discrete nature of CA is not different, qualitatively, from continuous CNN. We can introduce *continuous cellular automata* (CCA) as a generalization of CA in which each cell is not just black or white, for example, but instead can have any of a continuous range of possible levels of gray. A possible rule of a CCN may demand, for example, that the new gray level of each cell should be the average of its own gray level and those of its immediate neighbors. It turns out that in continuous cellular automata (CCA) simple rules of interaction can generate patterns of increasing *complexity*, *chaos*, and *randomness*, which are not essentially different to the behavior of discrete CA. Thus, they are useful approximations of the dynamics of systems which are determined by *partial differential equations* (PDE).

For the CNN paradigm, a neurobiological language delivers metaphorical illustrations of concepts, which are nevertheless mathematically defined and technically implemented. According to the dominant paradigms in life sciences today, a biological language mediates visions of future connections between bio- and computer technology. Mathematically, a CNN is defined by (1) a spatially discrete set of *continuous nonlinear dynamical systems* (*cells* or *neurons*) where information is processed into each cell via three independent variables (*input*, *threshold*, and *initial state*) and (2) a coupling law relating the relevant variables of each cell to all neighbor cells within a pre-described sphere of influence

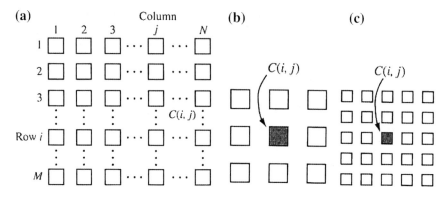

Fig. 4 Standard CNN with array (**a**), 3 × 3 and 5 × 5 neighborhoods (**b**, **c**)

(Chua 1998; Mainzer 2007). A standard CNN architecture consists of an $M \times N$ rectangular array of cells $C(i, j)$ with cartesian coordinates (i, j) with $i = 1, 2, ...,$ M and $j = 1, 2, ..., N$ (Fig. 4a). Fig. 4b–c shows examples of cellular spheres of influence as 3 × 3 and 5 × 5 neighborhoods. The dynamics of a cell's state are defined by a nonlinear differential equation (CNN state equation) with scalars for state x_{ij}, output y_{ij}, input u_{ij}, and threshold z_{ij}, and coefficients, called *synaptic weights*, modeling the intensity of synaptic connections of the cell $C(i, j)$ with the inputs (feedforward signals) from, and outputs (feedback signals) to, the neighbor cells $C(k, l)$. The CNN output equation connects the states of a cell with the outputs.

The majority of CNN applications use space-invariant standard CNNs with a cellular neighborhood of 3 × 3 cells and no variation of *synaptic weights* and *cellular thresholds* in the cellular space. A 3 × 3 sphere of influence at each node of the grid contains nine cells with eight neighbor cells and the cell in its center. In this case, the contributions of the output (*feedback*) and input (*feedforward*) weights can be reduced to two fixed 3 × 3 matrices which are called feedback (output) cloning template **A**, and feedforward (input) cloning template **B**. Thus, each CNN is uniquely defined by the two cloning templates **A**, **B**, and a threshold z, which consist of 3 × 3 + 3 × 3 + 1 = 19 real numbers. They can be ordered as a string of 19 *scalars* with a uniform threshold, nine feedforward and nine feedback synaptic weights. This string is called a CNN gene, because it completely determines the dynamics of the CNN. Consequently, the universe of all CNN genes is called the CNN genome. By analogy with the human genome project, steady progress can be made by isolating and analyzing various classes of CNN genes and their influences on CNN genomes.

A CNN program defined by a string of CNN genes is called a CNN chromosome. Every cellular automaton (CA) with binary states can be considered a CNN chromosome. In particular, Conway's Game-of-Life CA can be realized by a CNN chromosome. Since the *Game-of-Life* CA is a *universal Turing machine*, the corresponding Game-of-Life CNN is also a universal Turing machine. Thus, there is a universal CNN machine (CNN UM) which can simulate any particular CNN.

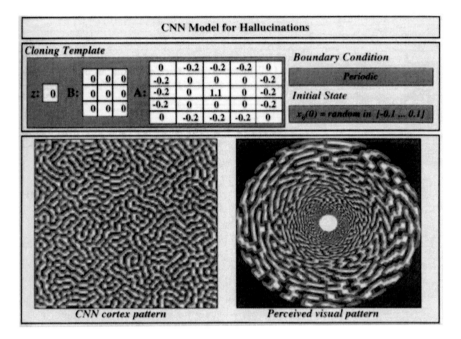

Fig. 5 CNN model for hallucinations

Applied to visual computing, the triplet {**A**, **B**, z} and its 19 real numbers can be considered as a CNN macro instruction on how to transform an input image into an output image. Simple examples are subclasses of CNNs with practical relevance such as the class $C(\mathbf{A}, \mathbf{B}, z)$ of space-invariant CNNs with excitatory and inhibitory synaptic weights, the zero-feedback (feedforward) class $C(0, \mathbf{B}, z)$ of CNNs without cellular feedback, the zero-input (autonomous) class $C(\mathbf{A}, 0, z)$ of CNNs without cellular input, and the uncoupled class $C(\mathbf{A}^0, \mathbf{B}, z)$ of CNNs without cellular coupling. In $\mathbf{A}^0$ all weights are zero, except for the weight of the cell at the center of the matrix. Their signal flow and system structure can be illustrated in diagrams that can easily be applied to electronic circuits as well as to typical living neurons.

Cellular Neural Networks (CNN) are optimal candidates to simulate *local neural interactions* of cells generating collective macro phenomena. A simple autonomous CNN was designed by using a template with local activation and lateral inhibition. It spontaneously generates a labyrinth pattern from random initial conditions. In the next step, the retina-cortical map is applied to the resulting stable pattern. Geometrically, a polar-coordinate point on the retina is mapped from the Cartesian point on the cortex, producing the perceived vision of a spiraling tunnel pattern (Fig. 5). The advantage of a CNN model is obvious: it can easily be programmed to a *CNN Universal Machine* (CNN-UM) chip which may be implemented into the living brain in future applications of neurosurgery.

Cellular Neural Networks (CNN)—with information processing in nanoseconds (with a standard design) and even at the speed of light (with optical technology)—

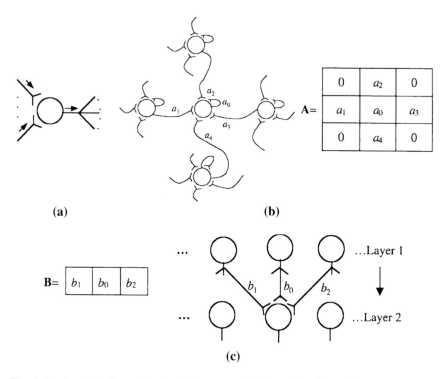

(a)   **(b)**

(c)

Fig. 6 CNN model of neurobionics with a neuron (**a**), **A**-template (**b**), and **B**-template (**c**)

seem to be optimal candidates for applications in *neurobionics*. Obviously, there are surprising similarities between CNN architectures and, for example, the visual pathway within the brain. An appropriate CNN approach is called the *Bionic Eye* (Chua and Roska 2002) which means a formal framework of vision models, combined and implemented on the CNN Universal Machine (CNN-UM). The analysis starts with a model of the receptive field organization in the retina and the visual pathway. Figure 6a shows a neuron with one axonal output with a branch to several other neurons and several dendritic inputs. The small gaps denote the synapses which are modeled by template elements. In Fig. 6b a neuron at the center receives recurrent inputs (outputs from its neighbors). Thus, the receptive field of a central neuron is modeled by a corresponding 3×3 **A**-template as its local sphere of influence. In Fig. 6c a part of a two-layer neuron network is shown with each layer as a one-dimensional representation of a two-dimensional grid. The neuron in the center of layer 2 receives dendritic inputs from the neighborhood in the input layer 1. The corresponding weights are modeled by a **B**-template.

Several neuroanatomic and neurophysiological models can be translated to CNN cloning templates. Length tuning, for example, means that certain neurons in the Lateral Geniculate Nucleus (LGN) and the visual cortex give a maximal response to an optimally oriented bar of a certain length. The response decreases or vanishes when the stimulus changes with an increase in the length of the bar.

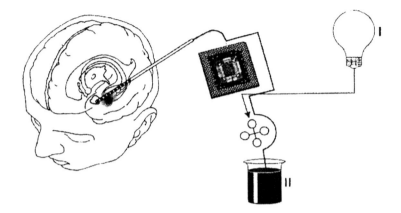

Fig. 7 CNN-UM chip for epileptic seizure prediction and prevention

A corresponding CNN model detects horizontal, vertical, and diagonal bars with length not longer than three pixels. Another function in the visual cortex is orientation selectivity, which can also be realized by an uncoupled CNN. Visual illusions which have been studied in *cognitive psychology* such as the arrowhead illusion can also be simulated by an uncoupled CNN. After introducing the "Lego" elements of the retina such as cells, synapses, and templates for receptive field organization, a simplified multi-layer CNN model of the retina can be designed and applied in *neurobionics*. In the end, the CNN-UM architecture allows the implementation of many spatio-temporal neuromorphic models. The same architecture of the universal machine can not only be used to mimic the retinas of different animals, such as of a frog, tiger salamander, rabbit, or eagle, but they can also be combined and optimized for technical applications. The combination of biological and artificial chips is no longer a science-fiction-like dream of cyborgs, but a technical reality, with inspiring prospects in the fields of *robotics* and *medicine* (Mainzer 2010).

In *epileptology*, clinical applications of CNN chips have already been envisaged (Tetzlaff 2002). The idea is to develop a miniaturized chip device for prediction and prevention of epileptic seizures. *Nonlinear time series analysis* techniques have been developed to characterize the typical EEG patterns of an epileptic seizure and to recognize the phase transitions leading to the epileptic neural states. These techniques mainly involve estimates of well-known criteria such as correlation dimension, Kolmogrov-Sinai entropy, Lyapunov exponents, measures for determinism, fractal similarity, etc. Implantable seizure prediction and prevention devices are already in use with *Parkinsonian patients*. In the case of epileptic processes, such a device would continuously monitor features extracted from the EEG, compute the probability of an impending seizure, and be provided with suitable prevention techniques. It should also possess both a high flexibility for tuning to individual patient patterns, and a high efficiency to enable estimation of these features in real-time. In addition, it should have a low energy consumption and should be small enough to be implemented in a miniaturized, implantable system. These requirements are optimally realized by *Cellular Neural Networks* (CNNs) with their massive parallel computing power, analogic information processing, and capacity for universal computation. Figure 7 shows a miniaturized chip device for seizure prediction and prevention. EEG-data are recorded from electrodes implanted near or within the epileptic area and fed to a time-series-based analysis system. The system extracts features of an impending seizure by a warning system (I) and supports an on-demand infusion of short-acting drugs to prevent the seizure (II).

Brains of Memristors?

On the horizon for future chip technology is the vision of *neuromorphic computers*, modeling the human brain with billions of neurons and synaptic connections. Brains are considered complex networks, with local cellular activities like those in CNNs and CAs. A technical unit modeling a living neuron with synapses, needs features for memory, digital circuitry, and a form of analog information processing. A strong candidate fulfilling all these requirements is the *memristor*, a new circuit element, which was suggested by one of the authors (Chua 1971), 40 years ago. Modern technology suggests that the memristor will bring a new wave of innovation in electronics, packing more bits into smaller volumes, and equipped with a kind of memory, preventing the loss of data.

The memristor device has generated immense interest among both device researchers and the memory-chip industry alike (Strukov et al. 2008). This interest was because of the high potential economic impact of the HP (Hewlett-Packard) breakthrough. Since the titanium-dioxide HP memristor could be scaled down to about 1 nm and is compatible with current IC technology, many industry experts are predicting that nano memristor devices would eventually replace both flash memories and DRAMS (*Dynamic Random Access Memory*). Indeed, a PC that requires no boot time, and which remembers all the data it was processing prior to disconnecting the power, could become a standard within a few years.

Memristor is an abbreviation for "memory resistor" und was predicted as the fourth missing circuit element with respect to the basic equations of electric circuits (Hayes 2011). These equations are defined for the four quantities *voltage* (v), *current* (i), *charge* (q), and *magnetic flux* (φ). Each equation determines a relationship between two of these variables. Historically, the oldest relation is Ohm's law $v = RI$, meaning that voltage is proportional to current. The constant of proportionality is given by the *resistance R*. If a current of I amperes flows through a resistance of R ohms, then the voltage with respect to the resistance is v volts. Geometrically, the graph of current versus voltage for a resistor is a straight line with slope R.

There are equations with different pairs of variables defining capacitors and inductors. Current and voltage are considered in terms of charge and flux. Thus, we get five equations with different pairings of the four variables v, I, q, and φ. But

Fig. 8 Four basic circuit
elements

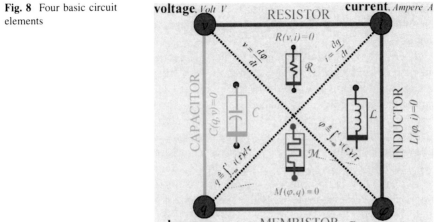

four objects taken two at a time provide six possible combinations. The missing
equation connects charge q and magnetic flux φ, thus determining the new element, joining the resistor, the capacitor, and the inductor. Obviously, the memristor was found by arguments of *symmetry* and *completeness*. This symmetry is
visualized in the diagram of Fig. 8.

The *resistor*, the *capacitor*, the *inductor*, and the *memristor* are all defined as
passive circuit elements, which must be distinguished from active devices, such as
transistors, which can amplify signals and inject power into circuits. Nevertheless,
the memristor has a characteristic ability, distinguishing it from the other passive
components: it is a nonlinear device. In a resistor, the relation between current and
voltage is a proportionality with a straight line of slope R. The memristor has a
graph in the form of a curve in the flux vs. charge plane, whose slope, called the
"memristance", varies from one point to another (Chua 2011).

A transistor is a three-terminal device with three connections to a circuit. It acts
as a switch or amplifier, with a voltage applied to one terminal controlling a
current flowing between the other two terminals. A memristor has only two terminals and cannot realize these functions. But memristors can be used to build
both memory and digital logic circuits. Metaphorically speaking, the memristor
has a built-in sense of history. A signal applied at one moment can affect another
signal that travels the same path later. The first signal exercises this control by
setting the internal state of the memristor to a high or low resistance.

Therefore, in a *neuromorphic computer*, memristors would not totally supplant
transistors, but would supplement them in memory functions and logic circuits.
Memristors could play the role of synapses. In biological neural networks, each
nerve cell communicates with other cells through thousands of synapses.
An important mechanism of learning occurs through adjustments to the strength of
the synaptic connections. In an artificial neural network, synapses must be small,
but effective structures. Memristors satisfy all the needed requirements. They

change their resistance in response to the currents that flow through them. This operation suggests a direct way of modeling the adjustment in the strength of synapses.

There are two qualitively distinct kinds of memristors, namely *locally passive memristors* and *locally active memristors*. The HP memristor is locally passive, because it does not require a power supply, and is said to be non-volatile. The potassium and sodium ion channels in the classic *Hodgkin-Huxley nerve membrane circuit model* can be considered locally-active memristors, powered by a sodium and a potassium pump, whose energy derives from ATP molecules. In contrast, synapses are locally passive memristors capable of retaining their synaptic efficacies over long periods of time without consuming any power.

Because our brains process information using only synapses and axons, it follows that circuits made of both types of memristors should also be able to emulate higher brain functions. The *long-term potentiation* (LTP) phenomenon associated with long-term memory can also be emulated by a memristor. Many *associative memory phenomena*, such as Pavlovian dog behavior, can be emulated by a memristor circuit. If brains are made of memristors, then we can expect that electronic circuits made of both locally passive and locally active memristors may someday emulate *human minds* (Mullins 2009). The key to this fundamental process is to uncover how *local activity could* lead to the *emergence of complex patterns* from a mass of homogeneous brain tissues. Formally, the local activity principle is realized in any cellular automaton. In neurons and memristors, the local activity principle is not only a formal model, but biological and technical reality.

References

U. Alon, Biological networks: the tinkerer as an engineer. Science **301**, 1866–1867 (2003)
U. Alon, *An Introduction to Systems Biology Design Principles of Biological Circuits* (Chapman & Hall/CRC, London, 2006)
L.O. Chua, Memristor–the missing circuit element. IEEE Trans. Circuit Theory **18**, 507–519 (1971)
L.O. Chua, *CNN: A Paradigm for Complexity* (World Scientific, Singapore, 1998)
L.O. Chua, Resistance switching memories are memristors. Appl. Phys. A **102**(4), 765–783 (2011)
L.O. Chua, T. Roska, *Cellular Neural Networks and Visual Computing: Foundations and Applications* (Cambridge University Press, Cambridge, 2002)
M. Creutz, Cellular automata and self-organized criticality. in *Some New Directions in Science on Computers*, ed. by G. Bhanot, S. Chen, P. Seiden. (Singapore, World Scientific, 1997), pp. 147–169
M. Gardner, The fantastic combinations of John Conway's new solitaire game of life. Sci. Am. **223**, 120–123 (1970)
M. Gardner, Mathematical games: on cellular automata, self-reproduction, the Garden of Eden, and the game "Life". Sci. Am. **224**(2), 112–117 (1971)
G. Gerisch, B. Hess, Cyclic-AMP-controlled oscillations in suspended dictyostelium cells: their relation to morphogenetic cell interactions. Proc. Natl. Acad. Sci. **71**, 2118 (1974)
H. Haken, A. Mikhailov (eds.), *Interdisciplinary Approaches to Nonlinear Complex Systems* (Springer, New York, 1993)

B. Hayes, The memristor. Am. Sci. **9**(2), 106–110 (2011)

Y. Kayama, Complex networks derived from cellular automata (Cornell University arxiv.1009.4509v1, 2010)

A. Kriete, R. Eils (eds.), *Computational System Biology* (Elsevier, Amsterdam, 2006)

K. Mainzer, Cellular Neural Networks and visual computing. Int. J. Bifurc. Chaos **13**(1), 1–6 (2003)

K. Mainzer, *Thinking in Complexity. The Computational Dynamics of Matter, Mind, and Mankind*, 5th edn. (Springer, Berlin, 2007)

K. Mainzer, *Leben als Maschine? Von der Systembiologie zur Robotik und Künstlichen Intelligenz* (Paderborn, Mentis, 2010)

J. Mullins, Memristor minds: the future of artificial intelligence. New Scientist 7 (2009)

D.B. Strukov, G.S. Snider, R. Duncan, D.R. Stewart, R.S. Williams, The missing memristor found. Nature **453**, 80–83 (2008)

R. Tetzlaff (ed.), *Cellular Neural Networks and their Applications* (World Scientific, Singapore, 2002)

P. Topa, Network systems modelled by complex cellular automata paradigm. in *Cellular Automata-Simplicity behind Complexity*, ed. by A. Salcido. (InTech, 2011), pp. 259–274

A.M. Turing, The chemical basis of morphogenesis. Philos. Trans. R. Soc. Lond. Ser. B Biol. Sci. **237**(641), 37–72 (1952)

Chapter 9
Outlook: Is the Universe a Computer?

To answer this question, a deeper analysis of our cosmological models is necessary. In general, *principles of symmetry* play a central role in physics. The *invariance* and *covariance* properties of a system under specific symmetry transformations can either be related to the conservation laws of physics or be capable of establishing the structure of the fundamental physical interactions and forces. This is the most essential aspect of symmetry because it concerns the basic invariance principles of physics and the interactions themselves, and not just the properties of geometric figures (Mainzer 1996).

With respect to *quantum physics* (Audretsch and Mainzer 1996), classical deterministic models are only approximations. Therefore, classical deterministic cellular automata are only approximate models of physical reality, which is governed by the principles of quantum physics ('t Hooft et al. 1992). Quantum cellular automata (QCA) would be more adequate but, of course, not as easy to understand as the toy world of classical cellular automata. In Chap. 7, we introduced the basic principles for QCA. In principle, it is possible to transform the concepts of quantum systems into QCA. For quantum cosmology, the concept of QCA must even be extended to quantum field theory.

Quantum cosmology uses quantum field theory to suggest a unification theory of physical forces represented by laws of symmetry. After the successful unification of the electromagnetic and weak interactions, physicists are attempting to realize the "*big*" unification of electromagnetic, weak and strong forces and, in a last step, the "*superunification*" of all four forces. The symmetry of this last unification would be the "*holy grail*" of modern cosmology and physics. There are several superunification research strategies, such as supergravity and superstring theories. Mathematically they are described by extensions of richer *symmetry* ("*gauge*") *groups.*[1] On the other hand, the variety of elementary particles is generated by *spontaneous symmetry breaking*. The concept of gauge symmetry

---

[1] In quantum physics, all the properties of a system can be derived from the state or wave function associated with that system. Formally, a *phase transformation* of the wave function $\psi(x, t)$ can be written as $\psi(x, t) \rightarrow \psi'(x, t) = e^{i\alpha} \psi(x, t)$ where α is the parameter (or phase) of the transformation

K. Mainzer and L. Chua, *The Universe as Automaton*, SpringerBriefs
in Complexity, DOI: 10.1007/978-3-642-23477-4_9, © The Author(s) 2012

and symmetry breaking plays an immense role in cosmology. During cosmic expansion and the decline in temperature, the initially unified *supersymmetry* of all forces fractured into the subsymmetries of those forces with specific physical interactions, and new elementary particles were generated at critical points in the phase transitions, leading to more variety and complexity.

Thus, *spontaneous breaking of gauge symmetries* leads to the emergence of new matter, patterns, and structure. It is obviously a fundamental principle in nature (Frampton 2008; Mainzer 2005a). The Higgs mechanism is a well known candidate for explaining spontaneous symmetry breaking and the emergence of massive particles (Goldstone 1961; Higgs 1964). Even the beginning of cosmic expansion (the "Big Bang") is assumed to be initiated by a kind of spontaneous symmetry breaking with respect to an equilibrium state of the quantum vacuum (Hawking et al. 2008).

To link *quantum cosmology* with *cellular automata*, quantum field theory must be discretized. *Lattice field theory* is the study of discretized lattice models of quantum field theory (Giles and Thorn 1977). In this case, quantum field theory is mapped onto a space–time that has been discretized onto a lattice. Although most lattice field theories are not exactly solvable, they are extremely interesting because they can be studied by simulation on a computer. The method is particularly appealing for the quantization of a gauge theory. Lattice field theory keeps manifest gauge invariance, but sacrifices manifest Poincaré invariance (McGuigan 2003).

Therefore, the methods of lattice field theories can be applied to the quantization of cellular automata. There is already *quantization of cellular automata* referring to particular categories of elementary particles (bosons and fermions), spins, strings, and supersymmetries. *Bosons* are particles that obey Bose–Einstein statistics. When two bosons are interchanged, the wave function of the system is unchanged. *Fermions*, on the other hand, obey Fermi–Dirac statistics. According to Pauli's exclusion principle, two fermions cannot occupy the same quantum state as one another, leading to the typical rigid features of matter. Thus fermions are sometimes said to be the constituents of matter (such as electrons and quarks), while bosons are said to be the particles transmitting interactions (such as gauge bosons and the Higgs boson) or radiation (e.g., the photon). All observed bosons have integer spin, as opposed to fermions, which have half-integer spin.

Footnote 1 (continued)

(Mainzer 2005b). If α is *constant*, i.e., the same for all points in space–time, the equation expresses the fact that once a phase convention has been made at a given point in space–time, the same convention must be adopted at all other points. This is an example of a global transformation applied to the field $\psi(x, t)$. If $\alpha = \alpha(x, t)$ is a *function of space and time*, then such a transformation will not leave any equation of $\psi(x, t)$ with space or time derivatives as invariant. This is, in particular, true for the Schrödinger equation or any relativistic wave equation for a free particle. To satisfy the *invariance under a local phase transformation* it is necessary to modify the equations in some way, which will no longer describe a free particle. Such modifications introduce additional terms, which describe the interaction of the particle with external fields and thereby generate the dynamics. That is the *gauge principle* or *principle of local symmetry*, according to which the interactions are determined by invariance under local symmetry (phase or gauge) transformations (Frampton 2008).

Cellular automata consist of a row of cells, the states of which can be updated at the next time step according to a local rule and their neighbors. This procedure can be generalized to higher dimensions by placing the cellular state on a higher dimensional lattice. Reversible cellular automata are a subclass of cellular automata that exhibit physical behavior such as locality and microscopic reversibility. Particular *reversible* QCA can be related to *discretized bosonic and fermionic field equations*.

One of the most fascinating feature of classical automata is the emergence of complex cellular patterns from very simple rules. QCA are also capable of simulating complex systems, but which emerge from nondeterministic or probabilistic rules of interactions between nearest neighbors. As in the quantum world, there are also strange quantum features which cannot be realized in the classical world of automata. Remember that the *garden of Eden* configurations of classical cellular automata cannot be reached by classical evolution of an update rule. In QCA, such arrays of cells can be reached quantum mechanically through quantum tunnelling, although selection rules forbidding certain transitions will still be possible (McGuigan 2003).

Summing up all these insights, we are on the way to conceiving quantum systems as QCA. In any case, the *Zuse-Fredkin thesis* must be revisited with respect to quantum physics: Is the universe a quantum cellular automaton? The answer to this question depends on the digitization of physics. The question *"Is the Universe a computer"* leads to the question: How far is it possible to map the laws of physics onto computational digital quantum physics? (Deutsch 1985). Digitization is not only exciting for answering philosophical questions of the universe. *Digitization* is the key paradigm in modern research and technology. Nearly all kinds of research and technical innovations depend on computational modeling. The emerging complexity of nature and society cannot be handled without computers with increasing computational power and storage.

To make this complex computational world more understandable, cellular automata are an excellent instructional tool. This booklet has shown that many basic principles of the expanding universe and the evolution of life and brain can be illustrated with cellular automata. The emergence of new structures and patterns depends on phase transitions of complex dynamical systems in the quantum, molecular, cellular, organic, ecological, and societal worlds (Mainzer 2007). Cellular automata are recognized as an intuitive modeling paradigm for complex systems with many useful applications (Hoekstra et al. 2010). In cellular automata, extremely *simple local interactions* of cells lead to the *emergence of complex global structures*. This *local principle of activity* is also true in the world of complex systems with elementary particles, atoms, molecules, cells, organs, organisms, populations, and societies (Chua 1998). Although local interactions generate a complex variety of phenomena in the universe, they can be mathematically reduced to some fundamental laws of symmetry.

Symmetries play a key role in the physical world as well as in the universe of automata. In the philosophy of science, they have been considered *universal principles* of *Platonic truth* and *beauty*. The scientific search for symmetries reminds us of Parsifal's quest for the Holy Grail. The legend of Parsifal was written by the minnesinger Wolfram von Eschenbach (*c.* 1170–*c.* 1220). It may be a random accord of names that a "Wolfram" also wrote a "New Kind of Science" for cellular automata. In the nineteenth century, Richard Wagner composed his famous opera based on Wolfram's legend of Parsifal. In Wagner's interpretation, it is the quest of the "poor fool" Parsifal, searching for the Holy Grail. The question is still open whether the scientific search for a final symmetry or "world formula" will also be a "foolish" quest.

References

J. Audretsch, K. Mainzer (eds.), *Wieviele Leben hat Schrödingers Katze? Zur Physik und Philosophie der Quantenmechanik*, 2nd edn. (Spektrum Akademischer, Verlag, Heidelberg, 1996)

L.O. Chua, *CNN: A Paradigm for Complexity* (World Scientific, Singapore, 1998)

D. Deutsch, Quantum Theory, the Church-Turing Principle and the Universal Quantum Computer, Proc. R. Soc. Lond. A **400**, 97–117 (1985)

P.H. Frampton, *Gauge Field Theories*, 3rd edn. (Wiley-VCH, New York, 2008)

R. Giles, C. Thorn, Lattice approach to string theory. Phys. Rev. D **16**, 366 (1977)

J. Goldstone, Field theories with 'superconductor' solutions. N. Cimento **19**, 154–164 (1961)

S.W. Hawking, J.B. Hartle, T. Hertog, The no-boundary measure of the universe. Phys. Rev. Lett **100**, 201301 (2008)

P.W. Higgs, Broken symmetries, massless particles, and gauge fields. Phys. Lett. **12**, 132–133 (1964)

A.G. Hoekstra, J. Kroc, P.M.A. Sloot (eds.), *Simulating Complex Systems by Cellular Automata* (Springer, Berlin, 2010)

K. Mainzer, *Symmetries of Nature* (De Gruyter, New York, 1996) (German 1988: Symmetrien der Natur. De Gruyter: Berlin)

K. Mainzer, *Symmetry and Complexity: The Spirit and Beauty of Nonlinear Science* (World Scientific, Singapore, 2005a)

K. Mainzer, Symmetry and complexity in dynamical systems. Eur. Rev. Acad. Eur. **13**, 29–48 (2005b)

K. Mainzer, *Thinking in Complexity. The Computational Dynamics of Matter, Mind, and Mankind* (Springer, Berlin, 2007)

M. McGuigan, Quantum cellular automata from lattice field theories (2003), http://arxiv.org/ftp/quant-ph/papers/0307/0307176.pdf

G. 't Hooft, K. Isler, S. Kalitzin, Quantum field theoretic behavior of a deterministic cellular automaton. Nucl. Phys. B **386**, 495 (1992)